▪ Workbook ▪

HEATING, VENTILATING, AND AIR CONDITIONING FUNDAMENTALS

Raymond A. Havrella

El Camino College

PRENTICE HALL
Englewood Cliffs, New Jersey   Columbus, Ohio

Cover photo: © FPG International
Editor: Ed Francis
Production Editor: Stephen C. Robb
Text Designer: Tally Morgan, WordCrafters Editorial Services, Inc.
Cover Designer: Brian Deep
Production Buyer: Deidra M. Schwartz
Production Coordination: WordCrafters Editorial Services, Inc.

This book was set in Century Schoolbook by The Clarinda Company and was printed and
bound by Semline Inc., a Quebecor America Book Group Company. The cover was printed
by Phoenix Color Corp.

Printed in the United States of America

10 9 8 7 6 5 4 3 2 1

ISBN 0-13-138769-3

Prentice-Hall International (UK) Limited, *London*
Prentice-Hall of Australia Pty, Limited, *Sydney*
Prentice-Hall of Canada Inc., *Toronto*
Prentice-Hall Hispanoamericana, S.A., *Mexico*
Prentice-Hall of India Private Limited, *New Delhi*
Prentice-Hall of Japan, Inc., *Tokyo*
Simon & Schuster Asia Pte. Ltd., *Singapore*
Editora Prentice-Hall do Brasil, Ltda., *Rio de Janeiro*

CONTENTS

TO THE TEACHER

This workbook is to be used in conjunction with *Heating, Ventilating, and Air Conditioning Fundamentals*. It can also be used as a supplement to any HVAC text. Each unit of this workbook provides lab activities corresponding to material in the textbook. The short objective test following each unit and the students' continually updated Progress Charts will enable you to assess each student's progress.

The Introduction explains the workbook's format, discusses the use of metrics and U.S. customary measurements, and presents—in great detail—important safety instructions. At the very beginning of the program the instructor should demonstrate the correct way to perform potentially hazardous tasks. Students should not begin Unit 1 performance tasks until they have passed the safety test given at the end of the Introduction.

Measurable performance objectives precede each unit. Before assigning work to a student or a group, which should not exceed four to a work station, demonstrate the proper method of performing the various tasks. By demonstrating, you will reinforce the students' knowledge of the proper procedures and tools needed to complete the job. In addition, define your criteria for evaluating the performance objective.

The Student Progress Chart on pages viii–x is to be filled in by the students as a record of their progress. By comparing each student's score with the highest possible score, you and the student can determine if more work must be done to master the skills needed to satisfactorily perform the jobs related to the reviewed material.

Students may be given extra credit for doing outside reading assignments or writing a detailed report on potential service problems relating to the reviewed material. Additional assignments such as removal and replacement of various system components of a work station can be given to fast workers; however, extra credit should be given only for work done in a manner accepted by the trade.

A basic curriculum might be two hours of lab work for each hour of lecture. The textbook and corresponding workbook can be used over one or two semesters—one chapter assigned for every one or two weeks.

Together the textbook and workbook provide material required for a number of short courses in heating, ventilating, and air conditioning. For example, you may choose to give an introductory course in solar energy, air balancing, heating, or basic refrigeration. The course you select would determine what chapters you would delete and the proper sequence of chapters to follow.

TO THE STUDENT

Heating, ventilating, and air conditioning is the fastest growing field among the construction trades. The industry is currently experiencing an explosion in new technology. Consequently, there is no better time than the present to get started in a challenging, well-paid career.

The HVAC industry is divided into three categories: residential, commerical, and industrial processing. By mastering the performance tasks in this workbook you will have a head start in becoming a qualified residential service technician. You will also have a working knowledge of many basic skills that should qualify you for an entry-level position as an HVAC mechanic servicing commerical and industrial equipment.

Review the textbook material that relates to the assigned perfromance tasks before doing the various jobs. The workbook information preceding the performance tasks will be extremely helpful if the text is not handy.

Your instructor will explain what is to be accomplished with each job. At the same time you will be given the conditions for completing the task and the method for evaluating your performance. Consult the following Student Progress Chart.

The short objective tests that follow each unit should reinforce the classroom lecture material. Select the correct response and review in your own mind why the other possible answers are incorrect. You should then reread and review the text material to increase your retention of the important information.

Student Progress Chart

JOB SHEET	TEXTBOOK REVIEW	ALLOTTED TIME	COMPLETION DATE	SCORE	EXTRA CREDIT

JOB SHEET	TEXTBOOK REVIEW	ALLOTTED TIME	COMPLETION DATE	SCORE	EXTRA CREDIT

JOB SHEET	TEXTBOOK REVIEW	ALLOTTED TIME	COMPLETION DATE	SCORE	EXTRA CREDIT

INTRODUCTION

This workbook is designed to help you learn the entry-level tasks that relate to heating, ventilating, and air conditioning. The units in this manual are systematically arranged to provide a logical sequence of learning experiences. Each unit contains several performance objectives. A student who successfully completes these performance tasks will acquire the necessary skills in the installation, design, maintenance, and servicing of heating, ventilating, and air-conditioning equipment.

Although the workbook is designed to accompany the text, *Heating, Ventilating, and Air Conditioning Fundamentals*, it also can be used as a lab supplement to any heating, ventilating, and air conditioning (HVAC) classroom textbook that contains the required theoretical background. The more theory an HVAC mechanic learns, the better troubleshooter he or she will become. Without theoretical knowledge—how and why a unit performs—one can only expect to be a parts changer, and hope for the best.

The job tasks at the end of each unit are designed to fulfill performance objectives. To help you complete each job, some technical information precedes each performance task. The technical information includes illustrations that demonstrate the proper methods and procedures for completing the various jobs.

■ HOW TO USE THIS BOOK

Before attempting the hands-on performance tasks, read the information on safety. You should be able to answer all the questions on the safety test correctly. The lab instructor will most likely present additional information, such as the location of fire extinguishers and other specific safety tips in your lab area.

Begin each unit by skimming through the pages and observing the headings and illustrations. Doing so will tell you what textbook information to review before attempting the unit performance tasks.

Next read each of the performance objectives listed on the job task sheets to find out what you must attempt to do and under what conditions. Be sure to consult the suggested reference materials if you require additional information. You can record your progress on the chart provided at the beginning of this book.

Each unit concludes with a written evaluation—multiple choice and essay questions. Read each test item, but do not answer any of the problems until you have completed all the jobs for that particular unit.

At this point, you should know the related text information, as well as what jobs you will be completing. Now, carefully read the introductory information and complete the procedural tasks.

You should not begin a new unit until you have successfully completed the written and manipulative tasks of the preceding unit.

■ THE METRIC SYSTEM

On December 23, 1975, President Ford signed into law the Metric Conversion Act of 1975. Since then the United States has attempted to convert from the U.S. customary units to the metric, or SI system *(Système International d'Unité)*. The decision to convert came about after many years of debate.

Because science, education, and industry now feel that our customary system of measurement is comfortable but costly, they are committed to make the necessary conversion as rapidly as possible. The customary system is fast becoming obsolete in world affairs, as well as being an economic and political handicap. But the size and diversity of American products make a massive changeover difficult and expensive. The changeover has therefore focused on cost and consumer convenience. However, there are still many troublesome questions—for example, how will certain products be converted? Consequently, when appropriate, both measurements are included in this book. Moreover, SI equivalents are subject to change.

The National Bureau of Standards publishes basic metric information and recommendations (see Table 1).

The metric system is simple and well suited for scientific research. The user need only multiply or divide by the basic units of 10 and its powers. The base-10, or decimal, system is the metric system. (Note the approximate conversion factors shown in Table 2.)

What are missing in Table 2 are the measuring units for pressure: the pascal (Pa); and the energy conversion unit, joule (J). (See Table 3.)

Because different reference points are used, it is difficult to convert air-conditioning charts to SI metric equivalents. For example, the British pressure-enthalpy chart for refrigerants uses 40°F as a reference point for enthalpy measurement; the SI metric chart measures from absolute zero (the Kelvin unit).

From a mechanic's point of view, you must use the available tools to make the necessary conversions. You can get the job done with the tools scaled in either system of measurement, but a more accurate measurement can be taken with SI metrics.

■ SAFETY

On December 29, 1970, the Congress of the United States passed the Williams-Steiger Occupational Safety and Health Act (OSHA). OSHA provides for job safety and health protection for all workers through the promotion of safe and healthful working conditions throughout the United States. Copies of the act, specific OSHA safety and health standards, and other applicable regulations may be obtained from the nearest OSHA regional office.

Safety is an important consideration on a job, especially when working with potentially dangerous tools, machines, and refrigerants. It is wise to begin your safe habits at your school lab. Those who do not practice safety in the school lab will hurt themselves and, most likely, be cited at a later date by an OSHA Compliance Safety and Health Officer. The following information details how to think and act safely while using various equipment on the job.

General
- Lifting heavy, awkward objects requires help. Do not strain and try to do it yourself.
- When lifting heavy objects, keep your back straight, and use your arms and legs.
- Avoid slips and falls by keeping the floor clean. Wipe up spilled liquids immediately.
- Oily rags are a fire hazard. They should be stored or discarded in a covered, metal container.
- Before working with movable machinery, remove all your jewelry and secure loose clothing and hair. It is very easy to become tangled in a machine's moving parts.
- Jewelry presents an additional hazard when one is working around electrical equipment. Be sure to remove watches, rings, etc., to prevent serious burns.
- Never attempt to work on a "live" electric circuit. Disconnect the power source before connecting or disconnecting electric connections.
- In addition to burns, solid-state components can be destroyed before you can react.
- Be sure to wear safety glasses when working on refrigeration equipment.
- A face shield must be worn while using the grinders. (Safety glasses will not protect your face if the wheel disintegrates.)

Table 1   Basic Metric Information

Here for all practical purposes is the metric system:

- Use the meter (m) for length
- Use the liter (l) for volume
- Use the gram (g) for mass

Common Prefixes Used

kilo	1 000 ⎫	More than one
hecto	100 ⎬	meter, liter,
deca	10 ⎭	gram
deci	0.1 ⎫	A part of *one*
centi	0.01 ⎬	meter, liter,
milli	0.001 ⎭	gram

Common Abbreviations

kilo	k
hecto	h
deca	da
deci	d
centi	c
milli	m

Seven Base Units in the Metric System

Unit	Name of Unit	Symbol
Length	meter	m
Time	second	s
Mass	kilogram	kg
Electric Current	Ampere	A
Temperature	Kelvin	K
Luminous Intensity	candela	cd

Supplementary Units

Plane angle	radian	rad
Solid angle	steradian	sr

The Metric System

Unit	Length of Number of Meters
kilometer (km)	1 000
hectometer (hm)	100
decameter (dam)	10
meter (m)	1
decimeter (dm)	0.1
centimeter (cm)	0.01
millimeter (mm)	0.001

Unit	Capacity Number of Liters
kiloliter (kL)	1 000
hectoliter (hL)	100
decaliter (daL)	10
liter (l)	1
deciliter (dL)	0.1
centiliter (cL)	0.01
milliliter (mL)	0.001

Unit	Weight of Number of Grams
metric ton (t)	1 000 000
kilogram (kg)	1 000
hectogram (hg)	100
decagram (dag)	10
gram (g)	1
decigram (dg)	0.1
centigram (cg)	0.01
milligram (mg)	0.001

Points to Remember:

- Although it sounds complicated, the metric system is really simpler than our antiquated English system.
- Anyone who can count to 10 can make sense of the system.
- All measurements go up or down in multiples of 10. If it goes up, you multiply; if it goes down you divide.

1 gram = Weight of 1 cubic centimeter of pure distilled water at a temperature of 4°

Note:

1 liter   = Volume occupied by 1 cubic decimeter

- A fire extinguisher should be provided for 3000 ft^2 (915 m^2) of protected building area. Travel distance from any point in the area must not exceed 100 ft (30.5 m).
- Electric fires should be extinguished with a nonconducting extinguishing agent such as carbon dioxide or dry powder (class-C extinguisher). Do not grab the hot wires in an effort to disconnect them.
- Water, foam, or multipurpose dry chemical extinguishing agents can be used on paper, wood, or cloth fires. *Do not* use water on

Table 2   How To Convert to Metric Equivalents
*Approximate Conversions to Metric Measures**

When You Know	Multiply by	To Find	When You Know	Multiply by	To Find
Length				*Length*	
inches (in)	25	milliliters (mm)	millimeters (mm)	0.04	inches (in)
inches (in)	2.5	centimeters (cm)	centimeters (cm)	0.4	inches (in)
feet (ft)	30	centimeters (cm)	centimeters (cm)	0.03	feet (ft)
feet (ft)	0.3	meters (m)	meters (m)	3.3	feet (ft)
yards (yd)	0.9	meters (m)	meters (m)	1.1	yards (yd)
miles (mi)	1.6	kilometers (km)	kilometers (km)	0.6	miles (mi)
Area				*Area*	
square inches (in²)	6.5	square centimeters (cm²)	square centimeters (cm²)	0.16	square inches (in²)
square feet (ft²)	0.09	square meters (m²)	square meters (m²)	10.8	square feet (ft²)
square yards (yd²)	0.8	square meters (m²)	square meters (m²)	1.2	square yards (yd²)
square miles (mi²)	2.6	square kilometers (km²)	square kilometers (km²)	0.4	square miles (mi²)
acres	0.4	hectares (ha)			
Mass (Weight)				*Mass (Weight)*	
ounces (oz.)	28	grams (g)	grams (g)	0.035	ounces (oz)
pounds (lb.)	0.45	kilograms (kg)	kilograms (kg)	2.2	pound (lb)
tons (2,000 lb.)	0.9	metric tons (t)	metric tons (t)	1.1	tons (2,000 lb)
Volume				*Volume*	
teaspoons (tsp)	5	milliliters (ml)	milliliters (ml)	0.06	teaspoons (tsp)
tablespoons (tbsp)	15	milliliters (ml)	milliliters (ml)	0.18	tablespoons (tbsp)
fluid ounces (fl oz)	30	milliliters (ml)	milliliters (ml)	0.03	fluid ounces (fl oz)
pints (pt)	0.47	liters (l)	liters (l)	2.1	pints (pt)
quarts (qt)	0.95	liters (l)	liters (l)	1.06	quarts (qt)
gallons (gal)	3.8	liters (l)	liters (l)	0.26	gallons (gal)
cubic feet (ft³)	0.03	cubic meters (m³)	cubic meters (m³)	35	cubic feet (ft³)
cubic yards (yd³)	0.76	cubic meters (m³)	cubic meters (m³)	1.3	cubic yards (yd³)
Temperature				*Temperature*	
Fahrenheit (°F)	5/9 (after subtracting 32)	Celsius (°C)	Celsius (°C)	9/5 (then add 32)	Fahrenheit (°F)

*Metric conversions are based on the approximate conversion factors as recommended by the National Bureau of Standards, U.S. Department of Commerce.

Table 3   Equivalents

One atmosphere	14.7 psia	101.3 K Pa
One inch of mercury (60°F)	0.491 psia	3.377×10^3 pascal
Enthalpy	Btu/lb	J/Kg
One foot-pound-force	1 ft-lb	1.356 joule
One British thermal unit	1 Btu	1.055×10^3 joule
One watt (1W)		One Joule Per Second (1/Js)
One British thermal units per hour	Btu/h	0.2929 Watt
1 Btu/ft²	0.271	1.136 joule/cm²

burning liquids or paints. Water will spread the fire.
- Use a tool for its intended use only. Use of the proper-size tool for the job helps to prevent damage or injury.
- Keep all tools sharp and in shape. Dull tools are dangerous and require a greater force to do the work.
- Tools with mushroomed heads or ones that are improperly dressed are not safe to use.
- Horseplay is dangerous in a shop.

Drill Press
- Always wear eye protection.
- Secure your work with a C clamp. (Your hands are inadequate.)

- Avoid using gloves and be careful of loose clothing or long hair.
- Before turning on the machine, remove the key from the chuck. Rotate the chuck by hand to see that it is not interfering with anything.

Parts Cleaner
- If a solvent splashes into your eyes, flush with water immediately and notify the instructor.
- Apply hand lotion to sore and red skin caused by cleaning solvents.
- Never use compressed air to blow solvent off clothing or skin. Compressed air will inject the solvent beneath the outer layers of skin.
- Do not place shop towels that have been used to wipe up solvent in your pockets. They will seep through your trousers and contact your skin.

Grinder and Wire Wheel
- Always wear a face shield when using grinders and wire wheels.
- Be sure that your workpiece is securely positioned against the grinding wheel or wire wheel.
- Avoid using gloves; and be careful of loose clothing or hair around the moving parts.
- Avoid conversation with others while grinding.

Welding Equipment
- Keep all equipment in good working condition.
- Do not oil pressure regulators or gauges.
- Use a friction lighter (striker) to light a torch.
- Point away from people and combustibles when lighting the torch.
- Oxygen in the presence of oil will burst into flame.
- Do not use oxygen to blow away dust.
- Place acetylene and oxygen drums in an upright position when in use.
- Secure tanks so that they cannot easily be tipped over. (Acetylene can explode with an impact pressure of 15 psi.)
- Check connections for gas tightness with a soap-bubble solution. (Acetylene connections use a left-handed thread so that oxygen and acetylene hoses cannot be interchanged.)

- Stand off to the side of the regulators when opening the tank valves.
- Never open the acetylene tank valve more than one-half turn.
- The oxygen tank must first be slowly cracked open because of its high pressure (2300 psi) and then back-seated to isolate the valve-stem packing.
- Purge the lines before lighting the torch.
- A malfunction due to a dirty tip, kinked hose, or faulty regulator can cause flashback. Flashback occurs when the flame suddenly disappears and a stream of soot is ejected from the tip, accompanied by a whistling sound. If this occurs, immediately turn off the torch valves and close the tank valves. Flashback can blow a regulator to pieces.
- Flashback results from improper maintenance and unsafe operating techniques. You can purchase check valves for flashback protection from the fuel-gas supplier.
- Always turn off the acetylene valve first.
- Replace or repair damaged hoses.
- Keep the flame away from concrete. (Intense heat may cause flying fragments.)
- Always weld or braze in a well-ventilated area. (Some silver brazing alloys contain cadmium, which gives off toxic vapors.)
- Keep a suitable fire extinguisher close at hand.
- Wear the proper clothing and eye shields.
- Prest-O-Lite tanks contain acetylene gas and acetone to keep the acetylene gas stable. They must be in an upright position when in use or the acetylene and acetone may separate.

Refrigerants
- Always wear the proper clothing, shoes, and safety glasses when working with refrigerants. If refrigerants are spilled on the skin, evaporation can cause freezing and frostbite. If sprayed on your eye, it may even freeze your eyeball.
- Even high-boiling refrigerants which are good solvents can leave your skin dry and cracked.
- The recommended procedure to follow if liquid refrigerant hits your skin is as follows:
  1. Soak in lukewarm water for 10 to 15 min.
  2. Apply a light coat of ointment such as a petroleum jelly, mineral oil, or similar material that is handy.

3. Do not use a bandage unless the abrasion is exposed to rubbing or other contact.

4. See a doctor.

- Take special care when systems have undergone a motor burnout. Contaminated oil often contains hydrochloric and hydrofluoric acids that are harmful to the skin.
- Freon refrigerants decompose in the presence of an open flame or an electric heating element. They form acids and a poisonous gas called phosgene.
- Deliberate inhalation of Freon refrigerants to produce intoxication can be fatal.
- When a refrigerant leak occurs, make sure the area is well ventilated before attempting to braze refrigerant lines.
- Refrigerant cylinders must meet the United States Department of Transportation regulations on the type of metal container, wall thickness, and testing procedures.
- The safe-handling rules for refrigerant cylinders include the following:

    1. Open valves slowly.
    2. Replace outlet caps when finished.
    3. Never force connections.
    4. Do not tamper with safety devices.
    5. Do not alter cylinders.
    6. Do not drop, dent, or abuse cylinders.
    7. Protect from rusting during storage.

- Federal regulations prohibit cylinders from being liquid full at 130°F (54°C) to avoid the possibility of hydrostatic pressure build-up. To prevent this condition, never refill a charging cylinder over 80 percent of its liquid capacity.
- Never heat cylinders with live steam or an open flame. Heating with a torch weakens the cylinder in the small concentrated area where the heat is applied.
- To speed up the flow of refrigerant during the charging process, place the refrigerant cylinder in a bucket of warm water [not to exceed 125°F (52°C)].
- A person can be fined $10,000 and imprisoned up to 10 years for refilling a throwaway or disposable refrigerant cylinder.
- Use a recovery unit. Don't vent refrigerants.

Monthly Safety Inspection Report

Inspect your lab facilities each month. Then score each of the 20 items on the following safety checklist from 0 to 5. Circle the number that best describes and suits your lab. Zero is the lowest (the most dangerous) and five is the highest (the most safe). Which areas require corrective action?

SAFETY CHECKLIST

1. 0 1 2 3 4 5   Good housekeeping
2. 0 1 2 3 4 5   Material handling
3. 0 1 2 3 4 5   Material piling and storage
4. 0 1 2 3 4 5   Aisles and walkways
5. 0 1 2 3 4 5   Machinery and equipment
6. 0 1 2 3 4 5   Electrical and welding equipment
7. 0 1 2 3 4 5   Tools
8. 0 1 2 3 4 5   Ladders and stairs
9. 0 1 2 3 4 5   Floors, platforms, and railings
10. 0 1 2 3 4 5   Exits
11. 0 1 2 3 4 5   Lighting
12. 0 1 2 3 4 5   Ventilation
13. 0 1 2 3 4 5   Protective clothing and equipment
14. 0 1 2 3 4 5   Explosion hazards
15. 0 1 2 3 4 5   Unsafe practices
16. 0 1 2 3 4 5   First-aid facilities
17. 0 1 2 3 4 5   Washroom and locker room
18. 0 1 2 3 4 5   Drinking fountain
19. 0 1 2 3 4 5   Fire-fighting equipment
20. 0 1 2 3 4 5   Guards and safety devices

After completing your assessment of lab safety, compare your findings with other members of your class. You may be able to take measures to correct many of the problem areas. If you are not able to, you will be more aware of the potentially dangerous areas.

Personal Safety Test

Before you begin any lab work, you must be well versed on safety precautions in the use of all shop equipment. Now take the "Personal Safety Test."

Directions

For the following questions write a T to the left of each statement if it is *true* and an F if it is *false*.

■ PERSONAL SAFETY TEST

_____ 1. Oily rags should be placed in a covered container.

_____ 2. Dull tools are as safe to use as sharp ones.

_____ 3. Refrigerant cylinders should never be filled to more than 70 percent of their capacity.

_____ 4. A striker should always be used when lighting an oxyacetylene torch.

_____ 5. For tightening nuts, an open-end wrench is better to use than an adjustable wrench.

_____ 6. It is safe to wear rings and watches while working around air-conditioning equipment.

_____ 7. It is safe to use 20 psi of acetylene when welding.

_____ 8. Safety glasses with a tinted lens must be worn while silver brazing.

_____ 9. Most accidents are due to carelessness.

_____10. There is no need for using a face shield while grinding.

_____11. The power should be shut off while working on electric wiring.

_____12. It is safe to wear loose-fitting clothing while operating moving machinery.

_____13. Dirty tools are responsible for some accidents.

_____14. The first-aid kit in this shop is located in the tool crib.

_____15. Mushroom-headed tools can be dangerous.

_____16. Machinery rooms should be well ventilated.

_____17. Gasoline is a safe cleaning fluid.

_____18. When lifting heavy objects, it is safer to lift with the back than the legs.

_____19. Horseplay is dangerous in a shop.

_____20. You should always remove the chuck key and rotate the chuck by hand before turning on a drill press.

_____21. Safe working conditions exist in dirty shops.

_____22. A fire extinguisher must be within 100 yards (91.5 m) of travel from any point in the lab area.

_____23. Foam or dry powder must be used on electric fires.

_____24. For brazing copper, set the acetylene pressure regulator for 15 psi and the oxygen regulator for 5 psi.

_____25. The oxygen cylinder valve should not be opened more than one-half turn.

_____26. When shutting off the torch, the acetylene is turned off first.

_____27. Acetylene can safely be used at pressures up to 30 psi.

_____28. The oxygen cylinder valve should be turned out rapidly to blow out any dust particles lodged in the pressure regulator or torch butt.

_____29. Equipment having moving parts should have a belt guard.

_____30. Freon refrigerants in the presence of an open flame produce phosgene gas.

_____31. It is safe practice to wear gloves while using the grinder.

_____32. If solvent gets splashed into the eyes, you should flush with water and immediately notify the instructor.

_____33. Nonreturnable cylinders can be refilled with small amounts for conveniently servicing roof-top units.

_____34. If the skin is hit with liquid refrigerant and frostbite occurs, you should soak the burn in lukewarm water and apply a light coat of ointment.

_____35. Freon refrigerants are safe refrigerants that permit deep inhalation of vapors without harmful effects.

This is to certify that I have received instruction on safety precautions in the use of all shop equipment. I fully understand my duties and responsibilities when using air-conditioning and refrigeration lab equipment.

Signature: _____

Unit 1
BENDING AND JOINING TUBING

■ INTRODUCTION

A heating, ventilating, and air-conditioning (HVAC) service technician must be an expert at joining and bending tubing. Small leaks cannot be tolerated. Kinks in tubing increase friction loss and horsepower requirements. Moreover, if a job is not piped neatly, the installer is held responsible for any damages.

Successfully completing the job sheets at the end of this unit should equip you with entry-level job skills for various pipe trades. The Student Progress Chart at the beginning of this workbook provides guidelines for judging how well you performed each task.

The following technical information on copper tubing is taken from the *Copper Tube Handbook* with the permission of the Copper Development Association, Inc. This information, along with illustrations furnished by various tool manufacturers, will tell you how to bend and join copper tubing properly.

■ TUBING

Copper tubing for air-conditioning work is available in 50-ft (15.2-m) coils of soft drawn tubing that can be easily bent into position, and 20-ft (6.09-m) lengths of hard drawn tubing that must be fitted where bends are required (see Figure 1–1).

Although the United States is changing to the international standard of measurement (SI metrics), pipe sizing will still be given in nominal inches. To find the metric equivalent in meters, multiply the inches by 0.0254.

Copper tubing for air conditioning and refrigeration (ACR tube) is cleaned and its ends are capped to keep out dirt and moisture. Type L copper is generally used unless type K, which has a heavier wall thickness, or type M, which has a thin wall, are specified by the engineer. Also, the outside dimension (OD) is referred to for pipe size. See Table 1–1 for the physical characteristics.

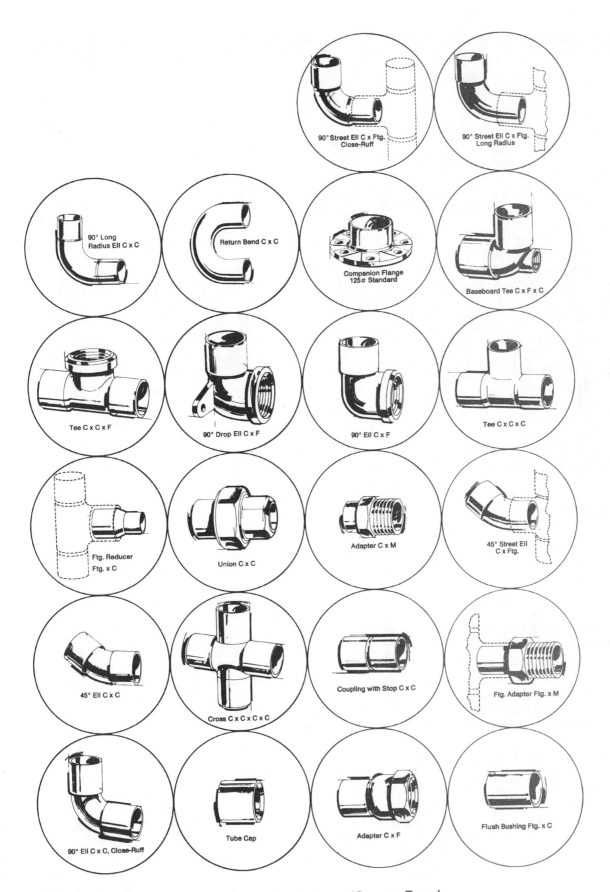

FIGURE 1–1 Copper fittings and adapters. *(Copper Development Association, Inc.)*

TABLE 1–1 Physical characteristics of copper tube.

	Nominal Dimensions, in			Calculated Values, Based on Nominal Dimensions			
Size, in	Outside Diameter	Inside Diameter	Wall Thickness	Cross-sectional Area of Bore, in²	External Surface, ft²/lin ft	Internal Surface, ft²/lin ft	lb/lin ft
1/8	0.125	0.065	0.030	0.00332	0.0327	0.0170	0.0347
3/16	0.188	0.128	0.030	0.0129	0.0492	0.0335	0.0577
1/4	0.250	0.190	0.030	0.0284	0.0655	0.0497	0.0804
5/16	0.312	0.248	0.032	0.0483	0.0817	0.0649	0.109
3/8	0.375	0.315	0.030	0.0780	0.0982	0.0821	0.126
3/8	0.375	0.311	0.032	0.0760	0.0982	0.0814	0.134
1/2	0.500	0.436	0.032	0.149	0.131	0.114	0.182
1/2	0.500	0.430	0.035	0.145	0.131	0.113	0.198
5/8	0.625	0.555	0.035	0.242	0.164	0.145	0.251
5/8	0.625	0.540	0.040	0.233	0.164	0.143	0.285
3/4	0.750	0.666	0.042	0.348	0.196	0.174	0.362
7/8	0.875	0.785	0.045	0.484	0.229	0.206	0.455
1 1/8	1.125	1.025	0.050	0.825	0.294	0.268	0.665
1 3/8	1.375	1.265	0.055	1.26	0.360	0.331	0.884
1 5/8	1.625	1.505	0.060	1.78	0.425	0.394	1.14
2 1/8	2.125	1.985	0.070	3.09	0.556	0.520	1.75
2 5/8	2.625	2.465	0.080	4.77	0.687	0.645	2.48
3 1/8	3.125	2.945	0.090	6.81	0.818	0.771	3.33
3 5/8	3.625	3.425	0.100	9.21	0.949	0.897	4.29
4 1/8	4.125	3.905	0.110	12.0	1.08	1.02	5.38

Note: Sizes shown in boldface type are available in annealed temper only; sizes shown in italics are available in hard temper only; all others are available in both. *Courtesy Copper Development Association, Inc.*

BENDING TUBING

Copper lines are used in most cases to connect the air-conditioning components. Properly bent, copper tubing will not collapse on the outside of the bend and will not buckle on the inside of the bend. Because copper is readily formed, expansion loops and other bends necessary in an assembly are quickly and simply made using the proper method and equipment.

Tools commonly used to bend tubing are shown in Figure 1–2. You must always slip the spring all the way onto the tubing. If a bend is made when the spring is not completely on the tubing, a permanent kink will be set in the bending spring and the tool will be damaged.

The lever-type bender is a little more difficult to use. However, if the manufacturer's instructions are followed, double offsets and bends will not present a problem.

EXPANSION LOOPS

Copper tube, like all piping material, expands and contracts when the temperature changes. Therefore, a copper-tube system subjected to excessive temperature changes tends to buckle or bend when it expands unless compensation is built into the system. The rate of expansion in inches per 100 ft can be determined from Figure 1–3.

Some type of expansion loop is required in the compressor discharge line or a leak will develop, usually a cracked elbow.

A residential hot-water heating system, with piping inside walls and through the attic, could develop serious trouble if expansion loops are not included in the piping. Every time the hot-water boiler comes on, the piping will expand with the rise in water temperature and the sound will be like mice playing in the attic.

■ **FIGURE 1–2**   Bending spring and mechanical bender.

Figure 1–4 shows three types of expansion loops. The developed length of the expansion offsets can be determined from Table 1–2 for configurations *A* and *B* and from Table 1–3 for the screwed pipe, illustration *C*.

■ FLARED JOINTS

Impact or screw-type tools (Figure 1–5) are used for flaring tube. The procedure for impact flaring is as follows (photos 1, 2, and 3):

1. Cut tube to desired length with a tube cutter.
2. Use a reamer (attached to the tube cutter) to remove the small burr on the end of the tubing.
3. Use a flat file to square off the end of the tube.
4. Slip the flare nut over the end of the tube.
5. Insert impact flaring tool into the tube (photo 1).
6. Drive the flaring tool by hammer strokes,

expanding the end of the tube to the desired flare. This requires a few moderately light strokes (photo 2).

7. Assemble the joint by placing the fitting squarely against the flare. Engage the coupling nut with the fitting threads (photo 3). Tighten with two wrenches, one on the nut and one on the fitting.

When using the screw-type flaring tool, the procedure is as follows (photos 4, 5, and 6):

1. Follow steps 1 to 4 given for impact flaring.
2. Clamp the tube into the flaring block so that the end of the tube is slightly above the face of the block (photo 4).
3. Place the yoke of the flaring tool on the block so that the beveled end of the compressor cone is over the tube end.
4. Turn the compressor screw down firmly, forming the flare between the chamber in the flaring block and the beveled compressor cone (photo 5).

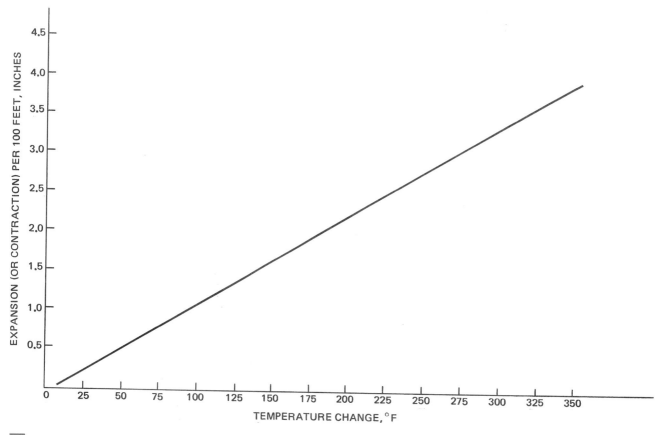

FIGURE 1–3 Expansion (per 100 ft) vs. temperature change for copper tube.

5.   Remove the flaring tool. The joint can be assembled as in step 7 for impact flaring (photo 6).

The screw-type flaring tool can also be used to reround tubing found to be slightly out of round.

SOLDERS

Solder joints depend on capillary action to draw the molten solder into the gap between the fitting and the tube. Capillary action can be described as a funneling effect, similar to water being swirled down into a drain. Flux, applied first, acts as a cleaning and wetting agent (fluidic flow permitting tube and fitting to bond together). When the flux is properly applied, it permits uniform spreading of the molten solder over the surfaces to be soldered. Capillary action is most effective when the space between the faces to be joined is between 0.002 and 0.005 in (0.05 to 0.127 m) in diametral clearance.

Moreover, to ensure capillary action, you must properly apply heat to the pipe first, so that the pipe expands inside the fitting and provides the close tolerance (0.05 to 0.127 m). Second, after heating the pipe directly above the joint to be soldered, you must apply the torch to the fitting below the joint. When the fitting and pipe reach the solder melting point, solder is then applied to the fitting. The solder is applied to the side opposite the flame contact, and only until the solder encircles the joint. 95–5 tin-antimony solder should be used. Sometimes 95–5 solder is specified on refrigerant lines. For temperatures exceeding 250°F (121.1°C), or where highest joint strength is required, brazing filler metals should be used.

FLUXES

Soldering flux functions to remove residual traces of oxides, to promote wetting, and to protect the surfaces from oxidation while heat is being applied. The flux should be applied to clean surfaces and only enough to lightly coat the areas to be joined should be used.

The fluxes best suited for 95–5 solder are mildly corrosive liquid or petroleum-based pastes containing chlorides of zinc and ammonium.

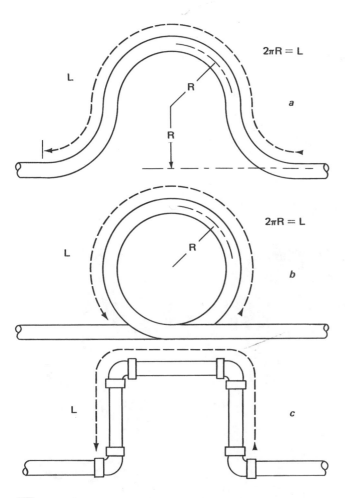

$2\pi R = L$

a

$2\pi R = L$

b

c

■ FIGURE 1–4 Configurations for expansions, loops, and offsets. *(Copper Development Association, Inc.)*

Figure 1–6 illustrates the behavior of flux during the brazing cycle. If Copper-Phos (Sil-Fos) filler material is used, flux is necessary for brazing copper pipe to a dissimilar material (brass or steel). However, flux is not needed when brazing copper to copper.

The silver solders listed under the BAg series (Figure 1–6) require a special flux for all applications. The BAg-1 and BAg-2 silver solders contain cadmium. Heating when brazing can produce toxic fumes; therefore, you should avoid breathing fumes and adequate ventilation must be provided when BAg-1 and BAg-2 are used.

The special flux used for silver brazing provides a temperature indicator, as shown in Figure 1–6. This is very helpful, because overheating weakens the joint. The strongest joint is made when the minimum heat is applied to the pipe and fitting to allow the filler material to flow and bond the connection together.

■ JOINTS

There are twelve simple steps to make a solder joint:

1. Measure length of tube.
2. Clean tube square.
3. Ream cut end.
4. Clean tube end.
5. Clean fitting socket.
6. Apply flux to tube end.
7. Apply flux to fitting socket.
8. Assemble.
9. Remove excess flux.

Table 1-2   Radii of Coiled Expansion Loops

Expected Expansion in	Radius R for Tube Size Shown, in												
	¼	⅜	½	¾	1	1¼	1½	2	2½	3	3½	4	5
½	6	7	8	9	11	12	13	15	16	18	19	20	23
1	9	10	11	13	15	17	18	21	23	25	27	29	32
1½	11	12	14	16	18	20	22	25	28	30	33	35	39
2	12	14	16	19	21	23	25	29	32	35	38	41	45
2½	14	16	18	21	24	26	29	33	36	40	43	45	51
3	15	17	19	23	26	29	31	36	40	43	47	50	55
3½	16	19	21	25	28	31	34	39	43	47	50	54	60
4	17	20	22	26	30	33	36	41	46	50	54	57	64

Courtesy Copper Development Association, Inc.

Table 1-3 Developed Length of Expansion Offsets

Expected Expansion in	Radius R for Tube Size Shown, in												
	¼	⅜	½	¾	1	1¼	1½	2	2½	3	3½	4	5
½	38	44	50	59	67	74	80	91	102	111	120	128	142
1	54	63	70	83	94	104	113	129	144	157	169	180	201
1½	66	77	86	101	115	127	138	158	176	191	206	220	245
2	77	89	99	117	133	147	160	183	203	222	239	255	284
2½	86	99	111	131	149	165	179	205	227	248	267	285	318
3	94	109	122	143	163	180	196	224	249	272	293	312	348
3½	102	117	131	155	176	195	212	242	269	293	316	337	376
4	109	126	140	166	188	208	226	259	288	314	338	361	402

Courtesy Copper Development Association, Inc.

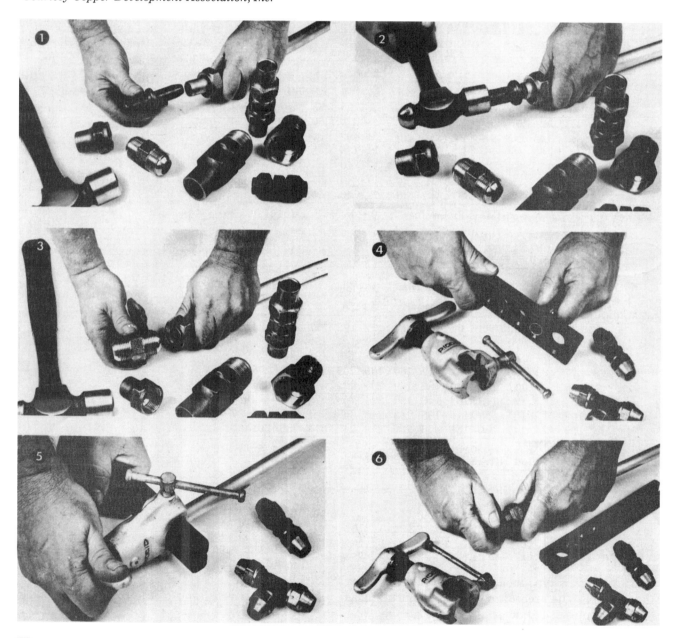

FIGURE 1–5   Procedure for impact flaring. *(Copper Development Association, Inc.)*

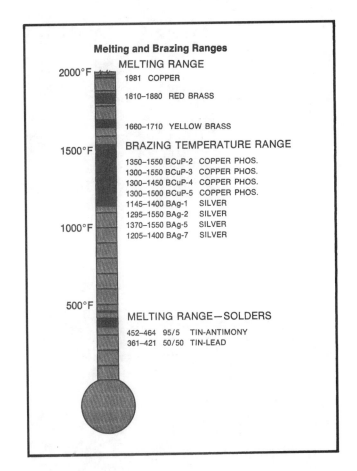

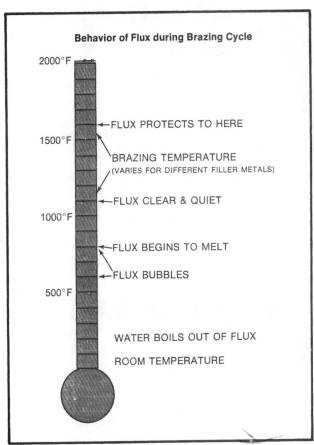

■ **FIGURE 1–6**   Brazing parameters. *(Copper Development Association, Inc.)*

10. Apply heat.
11. Apply solder.
12. Allow joint to cool.

Figure 1–7 illustrates these major steps.

■ BRAZED JOINTS

Strong leak-tight brazed connections for copper tube may be made by brazing with filler metals which melt at temperatures in the range between 1100 and 1500°F (593.3° to 815.5°C). Brazing filler metals are sometimes referred to as "hard solders" or "silver solders." These terms are confusing and should be avoided.

Brazing filler metals suitable for joining copper tube are of two classes: (1) alloys containing 30 to 60 percent silver (the BAg series) and (2) copper alloys which contain phosphorus (the BCa series). Refer to Figure 1–6 for brazing temperature ranges.

■ APPLY HEAT AND BRAZING

The procedure for brazing copper tubing (Figure 1–8) is as follows:

1. Clean the tubing and apply flux with a brush before assembling.
2. Assemble the joint by inserting the tube into the socket hard against the stop, and turn if possible. The assembly should be firmly supported so that it will remain in alignment during the brazing operation.
3. Apply heat to the parts to be joined, preferably with an oxyacetylene flame. Air acetylene is sometimes used for smaller sizes. A slight reducing (excess fuel) flame should be used, with a feather on the inner blue cone; the outer portion of the flame should be white. Heat the tube by beginning about 1 in from the edge of the fitting and sweeping the flame

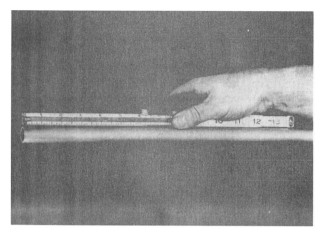
1. Measuring

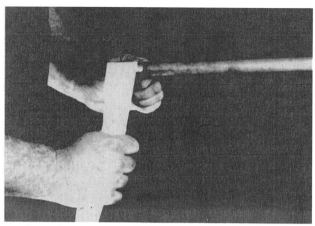
4. Cleaning tube end

2. Cutting

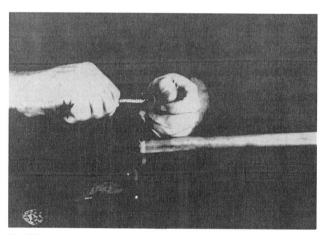
5. Cleaning fitting socket

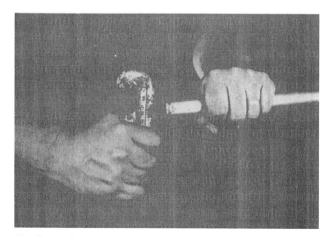
3. Reaming

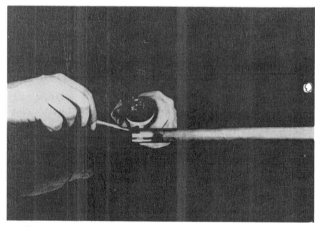
6. Fluxing tube end

FIGURE 1–7 Twelve steps in making a solder joint. *(Copper Development Association, Inc.)*

7. Fluxing fitting socket

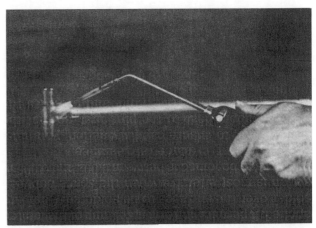

10. Heating the assembly

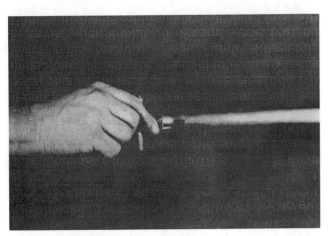

8. Assembling fitting and tube

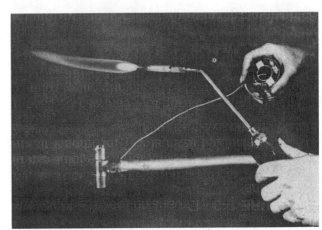

11. Applying solder

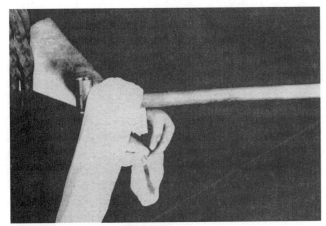

9. Removing excess flux

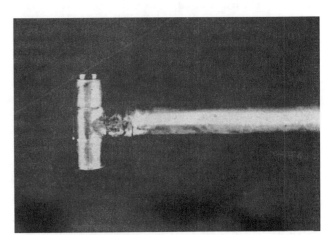

12. The finished joint

FIGURE 1-7 *Continued.*

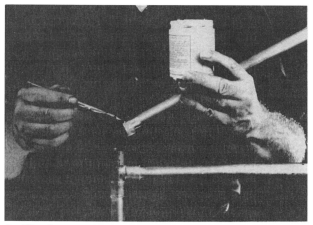

I. Fluxing

3. Heating small tube

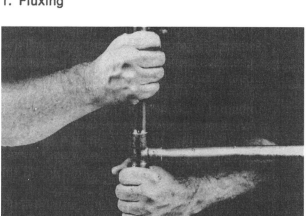

2. Assembling

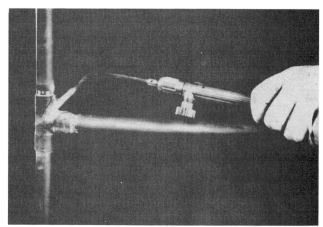

4. Heating small fitting

■ **FIGURE 1–8**   Four steps in brazing. *(Copper Development Association, Inc.)*

around the tube in short strokes at right angles to the axis of the tube. It is very important that the flame be in motion continuously and not remain at one point long enough to damage the tube. Use the flux as a guide. Continue heating the tube until the flux turns transparent, like water.

4. Switch the flame to the fitting at the base of the cup. Heat uniformly, sweeping the flame from the fitting to the tube until the flux on the fitting becomes quiet. The flame must be kept moving to avoid burning the tube or fitting.

5. Apply the brazing wire, rod, or strip at the point where the tube enters the socket of the fitting. When the proper temperature is reached, the filler metal will flow readily. If the joint is not hot enough to melt the filler metal, remove the rod and continue heating. When the joint is filled, a continuous strip of filler metal will be visible completely around the joint. Stop feeding as soon as the joint is filled.

For a 1-in tube and larger it may be difficult to bring the whole joint up to heat at one time. You will frequently want to use a multiple-tip torch to maintain the proper temperature over the large area. A mild preheating of the whole fitting is recommended for larger sizes. Heating can then proceed as outlined in the steps above. If you encounter difficulty in heating the entire joint to the desired temperature at one time, heat and braze a portion at a time, proceeding by segments progressively all around the joint.

■ HORIZONTAL AND VERTICAL JOINTS

When you make horizontal joints it is preferable to apply the filler metal at the bottom first, then at the two sides, and finally at the top, making sure the operations overlap. It does not matter where the start is made on vertical joints. If the opening of the socket is pointing down, you should be careful to avoid overheating the tube, because this may cause the brazing alloy to run down the outside of the tube. If this happens, take the heat away and allow the alloy to set. Then reheat the cup of the fitting to draw up the alloy.

■ REMOVING RESIDUE

After the brazing alloy has solidified, clean off flux residue with a wet brush or swab. Wrought fittings may be chilled quickly. However, allow cast fittings to cool naturally to some extent before wetting.

■ GENERAL HINTS AND SUGGESTIONS

If the filler metal fails to flow or has a tendency to ball up, this is an indication of oxidation on the metal surfaces or not enough heat on the parts to be joined. If work starts to oxidize during heating, this means that there is too little flux.

■ ALUMINUM TUBING

Because of lower price and availability, aluminum tubing is being used on many factory-assembled, packaged air-conditioning units. Some manufacturers are even using aluminum condensers. It takes more skill to repair and join aluminum tubing than it does to repair or join copper.

The three most common problems are repairing a leak, brazing aluminum tubing, and brazing aluminum tubing to copper. The simplest way to repair a leak is to use epoxy. But if the wrong epoxy is used, or if the manufacturer's instructions are not carefully followed, the epoxy will harden like glass and crack open once the metal is exposed to temperature changes.

To braze aluminum tubing, prepare the metal as described earlier for brazing copper. However, substitute All-State No. 31 aluminum brazing flux and use a No. 716 gas welding, aluminum brazing rod. The flux is a blue powder, but it can be applied in the same manner as copper brazing flux if a small amount is mixed with a few drops of water and made into a loose paste (use a baby-food jar to mix flux).

When brazing aluminum tubing you should also sweep the flame over a broad area of the tubing rather than directly above and below the joint. Run the flame parallel to the tubing and cover at least 3 in (about 75 mm) above and below the joint. If the flame is concentrated in one area, you will burn a hole in the tubing.

Furthermore, when the flux turns to a liquid, take the brazing rod and actually scratch the joint. Rubbing the brazing rod across the joint breaks up oxidation. As soon as the aluminum reaches the proper temperature the brazing rod will flow like water and leave a smooth fillet.

Follow the same procedure for repairing small pin-hole leaks as you followed for joining aluminum tubing. However, use a patch to repair larger holes.

Now comes the hard part: joining copper to aluminum. The task is made easier by first tinting the copper tubing with BAg-1 (45 percent silver) copper brazing wire. Apply All-State No. 31 flux to the copper tubing and to the cleaned aluminum tubing, and then braze the joint with an aluminum brazing rod.

Next, observe the instructor's demonstration and complete the following performance objectives.

■ JOB SHEET 1–1 ■

Name _____

Score _____ Date _____

PERFORMANCE OBJECTIVE
Given the proper tools, material, and instructions, make solder joints in vertical and horizontal positions. Solder should fill the fitting but not flow down the outside or inside of the pipe.

REFERENCE
Copper Tube Handbook, Copper Development Association, Inc.

EQUIPMENT
Air-acetylene torch kit, copper-sweat fittings, work station with a vise

TOOLS
Tube cutter, flat file, flux brush

SUPPLIES
Sandcloth, flux, ⅜-in-OD (9.5-mm) annealed copper tubing, 95–5 tin-antimony solder

JOB 1–1
Soft soldering sweat fittings

PROCEDURE
1. Measure and cut three 18-in (0.45-m) copper tubing pieces.
2. Ream and file the ends.
3. Clean the ends with sandcloth.
4. Apply flux, with brush, to tube ends and copper tee sockets.
5. Secure one piece of copper vertically in a bench vise.
6. Connect a run end of a tee to the secured pipe.
7. Assemble the two remaining copper tubes.
8. Remove excess flux.
9. Apply heat.
10. Apply solder.
11. Allow to cool.

■ JOB SHEET 1–2 ■

PERFORMANCE OBJECTIVE
Bend copper tubing with and without the aid of bending tools. Do not crimp the tubing. Use the material from Job 1–1.

REFERENCE
Copper Tube Handbook, Copper Development Association, Inc.

EQUIPMENT
Work bench

TOOLS
⅜-in lever-type bender, ⅜-in-OD (9.5-mm) bending spring, automatic locking steel tape

SUPPLIES
Completed project from Job 1–1

JOB 1–2
Bending copper tubing

PROCEDURE
1. Measure 3 in (76.2 mm) from face of copper tee on the branch leg.
2. Make a 90° bend in the run direction of the tee. Measurement from face of tee to end of bend should be 4.5 in (11.4 cm) (see Figure 1–9a).
3. Measure 6 in (15.2 cm) from end of branch pipe and make a 90° bend in the direction of the branch line leaving the tee (Figure 1–9b).
4. Slide an outside bending spring over one of the run connections and make a 360° loop as close to the tee as possible (Figure 1–9c).
5. Measure in 4 in (10.1 cm) from the end of the looped run.
6. With the lever bender make a 45° bend from the previous mark (Figure 1–9d).
7. Using your thumbs as a fulcrum, grasp the center of the remaining run line and make a 90° bend, down over the thumbs. A circular 2.5- to 2.75-in (63.5- to 69.8-mm) wooden disk could also be used (Figure 1–9e).

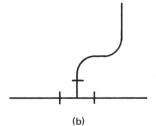

(a)

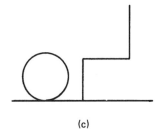

(b)

(c)

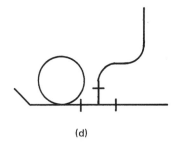

(d)

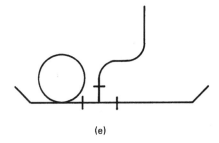

(e)

■ FIGURE 1–9

■ JOB SHEET 1-3 ■

Name _____
Score _____ Date _____

PERFORMANCE OBJECTIVE
Given the proper tools, make flare connections to allow the nut to
easily slip over the flare and large enough to allow sufficient sealing surface without the flare splitting.

REFERENCE
Copper Tube Handbook, Copper Development Association, Inc.

EQUIPMENT
Work station, Job 1–2 sketch

TOOLS
⅝-in swage tool, 8-oz ball-peen hammer, tubing cutter, flaring tool,
hammer-type flaring tool (½-in-ID tubing), flat file, 8- and 12-in adjustable wrench, tongue and groove pliers

SUPPLIES
⅝-in OD (15.8-mm) flare nut, ⅝-in OD flare plug, 6 in (15.2 cm) of
⅝-in hard drawn copper tubing, 6 in (15.2 cm) of ¼-in (6.35-mm)
copper tubing, ¼-in flare nut, completed project from Job 1–2

JOB 1–3
Flaring and swaging

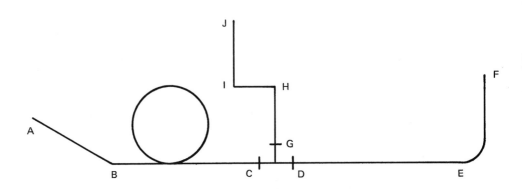

FIGURE 1–10 Complete sketch of Job 1–2.

PROCEDURE
1. Measure 2.75 in (69.8 mm) from *D* to *E* and cut tubing.
2. Ream the piece attached to the tee.
3. Swage the reamed and filed end.
4. Insert the *E* to *F* 90° elbow into the swage joint.
5. Crimp end *F* closed with tongue and grooved pliers.
6. Ream and file both ends of 6-in length of ⅝-in OD tubing.
7. Measure from the end of branch pipe *J* inward 1.5 in (38.1
    mm) and slide ⅝-in tubing over the branch pipe to the
    mark. Make a ⅝-in tube crimp joint. Slip ⅝-in nut over pipe
    and flare the end.

8. Tighten flare plug into ⅝-in flare nut.
9. Measure 2 in (about 50 mm) from end *A* and cut tubing.
10. Use ⅜-in swage tool on bend.
11. Insert *A* tube into *B* swage joint.
12. Insert ¼-in OD tubing 1 in (25.4 mm) into pipe *A*.
13. Attach ¼-in flare nut to pipe *A*.

■ JOB SHEET 1–4 ■

PERFORMANCE OBJECTIVE
Given the assembled project described in Job 1–3, silver braze the five connections, attach a drum of nitrogen to the ¼-in flare nut, and pressure-test the project to 100 psig.

REFERENCE
Copper Tube Handbook, Copper Development Association, Inc.

EQUIPMENT
Work station, with acetylene-oxygen welding rig

TOOLS
Tinted safety glasses, striker, pliers, flux brush, 8- and 10-in adjustable wrenches

SUPPLIES
Silver brazing rod and flux, sandcloth, nitrogen bottle and regulator, liquid soap bubbles

JOB 1–4
Silver brazing and pressure testing

PROCEDURE
1. Clean all fittings with sandcloth.
2. Apply flux and silver braze swage joints and crimped end.
3. Connect assembly to nitrogen regulator with the ¼-in flare nut.
4. Pressurize to 100 psig.
5. Apply soap bubbles to connections for a leak check.

■ JOB SHEET 1–5 ■

PERFORMANCE OBJECTIVE
Given an air-acetylene torch kit and aluminum tubing, braze an aluminum joint and also braze the aluminum pipe to the copper pipe project completed in Job 1–4. (Assembled project holds 150 psig pressure.)

EQUIPMENT
Work station, air-acetylene torch kit (No. 2 tip), Job 1–4

TOOLS
Safety glasses, striker, ⅜-in swage tool, tube cutter, flat file, flaring tool yoke, ball-peen hammer, two 8-in adjustable wrenches

SUPPLIES
BAg-1 (45 percent silver) copper brazing wire, No. 716 gas welding aluminum brazing rod, All-State No. 31 flux, silver solder flux, 12-in (0.3-m) piece of ⅜-in-OD aluminum tubing, sandcloth

JOB 1–5
Brazing aluminum tubing

PROCEDURE
1. Ream and file the ends of aluminum tubing.
2. Lightly coat the swaging tool with refrigerant oil.
3. Slip swage tool inside the tube to the coned section of the tool.
4. Place tubing in flaring tool yoke. The tubing must extend above the yoke, the distance measured from the end of the tubing with the inserted swage tool to the swage tool handle (approximately the outside diameter of the tubing plus ⅛ in).
5. Hammer the swaging tool into the tubing with sharp blows.
6. Repeat the procedure on the other end of the aluminum tubing.
7. Cut the swaged piece of aluminum tubing in half and rejoin it by placing the center-cut end into the swaged end of the second piece.
8. Clean swage joint with sandcloth and coat both pieces with flux.
9. Clamp tubing in a vise and heat joint with air-acetylene torch. When the flux turns to water, rub the aluminum brazing rod across the joint until rod flows and a fillet encircles the tubing.
10. Make a swage joint on a section of ⅜-in-OD copper from the project completed in Job 1–4.
11. Silver braze the copper swage joint and pull the joint apart before the silver braze (BAg-1 or comparable) cools and hardens.

12. Use All-State No. 31 flux on silver-brazed copper connections.
13. Clean and flux male and female ends of aluminum tubing.
14. Reconnect copper project by inserting aluminum tubing.
15. Use aluminum brazing rod with All-State No. 31 flux to make the two aluminum-to-copper connections.
16. Pressure-test to 150 psi using nitrogen.
17. Apply soap bubbles to connections for leak check.

■ MULTIPLE-CHOICE TEST ■

Name _____

Score _____ Date _____

DIRECTIONS
Circle the letter that best answers the following
multiple-choice questions.

1. Which of the following does not apply to
   ACR tubing?

   a. Soft drawn tubing comes in 50-ft coils
   b. Hard drawn is available in 20-ft lengths
   c. ACR tubing is sized by the inside diam-
      eter of the tubing
   d. ACR tubing comes cleaned and capped

2. Which of the metric equivalents is incorrect?

   a. ¼-in OD (2.5 mm)
   b. ⅜-in OD (9.5 mm)
   c. ½-in OD (12.7 mm)
   d. ⅝-in OD (15.8 mm)

3. Unless otherwise specified for ACR, use
   one of the following types of tube:

   a. K green color code
   b. L blue color code
   c. M red color code
   d. DWV yellow color code

4. An air-conditioning system requires an
   expansion loop:

   a. at the expansion metering device
   b. at the compressor suction line
   c. at the compressor discharge line
   d. at the liquid drain line

5. The following solder alloy contains anti-
   mony and can be used to connect refrigerant lines:

   a. 50–50   b. 60–40
   c. 95–5   d. Sil-Fos

6. Brazing alloys melt at temperatures beyond:

   a. 400°F
   b. 650°F
   c. 700°F
   d. 800°F

7. The silver brazing temperature range is
   between:

   a. 400 and 500°F
   b. 500 and 800°F
   c. 800 and 1100°F
   d. 1100 and 1500°F

8. The following copper-brazing alloy contains no silver:

   a. BCuP-2   b. BAg-1
   c. BAg-2   d. BAg-7

9. A male x female sweat fitting is called a:

   a. street ell
   b. baseboard ell
   c. close-ruff ell
   d. coupling

10. Which statement is false?

    a. 95–5 soldering flux contains chlorides
       of zinc and ammonium.
    b. 95–5 solder requires All-State No. 31
       flux.
    c. Copper-to-copper joints can be silver-
       brazed with copper-phosphorous
       alloys.
    d. Flux protects silver-brazed copper
       joints to 1600°F.

Unit 2

CONDENSING UNIT COMPONENTS

■ INTRODUCTION

A doctor uses a stethoscope to listen for various sounds produced within the body. Similarly, the refrigeration mechanic installs a gauge manifold to find out how the compressor body is pumping. Moreover, both use thermometers to check for malfunctions. Before an accurate diagnosis can be made, a mechanic must be able to identify the component parts of a system, know their function, and know how to check each component to see if it is performing properly.

The following tasks will involve identification of the compression-system components, as well as knowledge of pressure and temperature testing.

If the compressor is running but not functioning properly, gauges should be installed to check the high- and low-side pressures. Sometimes hermetically sealed units do not have service valves. If so, access valves must be installed before you can connect the gauges.

Figure 2–1 shows the front view of a remote condensing unit. The condenser is a wraparound design with a propeller-type fan. The fan pulls the air through the condenser and discharges it out the top of the unit. The rear view, with the access

panel removed, is shown in Figure 2–2. The cover for the electrical control box is also removed.

The compressor is *hermetically sealed*, which means that the motor and compressor are enclosed in a steel dome that cannot be unbolted for servicing. The motor and compressor are located in the center of the unit. The hermetically sealed unit does not use belts, pulleys, or shaft seals.

The upper-left corner of the compressor shows a copper stub that has been crimped and brazed. This is called a *process tube.* Some units are factory-charged through this line and no access valves are installed on the unit. This unit has a suction and a liquid-line service valve. It is located outside the unit at the lower left-hand corner. These valves eliminate the need for a process tube connection. However, if you need to isolate the compressor from the system, an access valve can be installed onto the process tube. This provides access to the compressor crankcase for evacuation, charging, or pressure readings. (See Figures 2–1 and 2–2.)

There are two main types of access valves. One type fits around the tubing. When the two halves are fastened together, a neoprene seal is pressed a one-piece access valve. The control valve can be removed and a seal cap can be placed over the piercing valve when the servicing is completed.

The second type of access valve is the Shrader valve. It is similar to a bicycle tire valve. The Schrader valve has a replaceable valve core that can be removed with the tool shown in the lower left-hand corner of Figure 2–4. To install it, drill a hole into the tubing and braze the valve to the copper line. Or, screw the pipe adapter into a female pipe thread. The valve will open when a charging hose, with an inserted depressor, is screwed onto the ¼-in male flare fitting.

A gauge manifold, such as the one shown in Figure 2–5, is connected to service valves with flexible hoses that have ¼-in female flare connections.

■ **FIGURE 2–1**   Remote condensing unit—front view. *(Square D Company)*

against the tubing (see Figure 2–3). Then, as the center screw is turned in, it pierces the tubing. The Watsco control valve has a screwdriver tip on the end of the stem to open or close the piercing valve. The knurled ring on the bottom of the control valve screws onto the piercing valve to make

(a)

(b)

(c)

■ **FIGURE 2–2**   Remote condensing unit—rear view. *(Square D Company)*

■ **FIGURE 2–3**   Tube piercing access valves. *(Watsco Products)*

■ **FIGURE 2–4** Schrader-type access valve. *(Watsco Products)*

■ **FIGURE 2–5** Gauge manifold. *(Imperial-Eastman Corp.—Valve and Fitting Division)*

The suction line connects to the low side (Figure 2–5). The discharge, or liquid line, connects to the high side. These two connections are always open from the gauge to the low side and high side ¼-in male flare connections, whether the manifold valves are open or closed.

The center flare connector, with a sight glass, connects to the manifold. The low-side gauge and hose can be opened to the manifold by turning counterclockwise the valve handle marked "low." Similarly, to open the high side, turn the valve marked "high" counterclockwise.

The two flare connectors that are screwed into the opposite ends of the manifold are Schrader valves. They open to the manifold when a hose with a depressor is screwed on; they automatically seat when the service hose is removed. These two connections are used when the system is being evacuated of air and noncondensibles.

The center tap is used to connect a refrigerant charging cylinder, a vacuum pump, or an oil pump when adding compressor oil (see Figure 2–6).

The temperature-testing equipment shown in Figure 2–7*a* is ideal for checking an AC cooling system. You can insert the dial-stem thermometer into a small duct access made with an awl. The stems are metal and can take quite a bit of abuse. And the dial scale is easy to read.

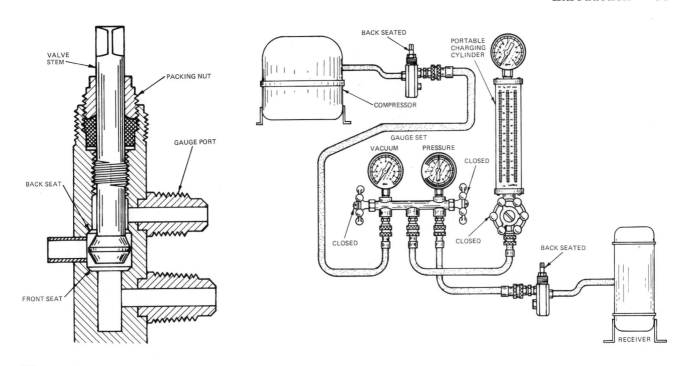

FIGURE 2–6   Gauge connections.

Another thermometer, the electronic temperature tester, has three leads. These leads can be remotely placed to sense temperatures in three different areas. A $\frac{3}{8}$ in × 1 in bulb thermister is placed on the end of each lead to sense the temperature. (See Figure 2–7b.)

You'll need a gauge manifold and temperature test equipment—similar to what we've just described—to perform the following jobs.

(a)   (b)

FIGURE 2–7   (a) Pocket (Robinair Products) and (b) electronic thermometers. (Weksler Instruments)

■ JOB SHEET 2–1 ■

Name _____

Score _____ Date _____

PERFORMANCE OBJECTIVE
Given a gauge manifold with charging hoses and hand tools, within 15 minutes locate service valves and properly install gauges on an operable unit.

REFERENCE
Heating, Ventilating, and Air Conditioning Fundamentals (Chapter 2); Figures 2–1 through 2–7

EQUIPMENT
Central air-conditioning unit, remote condensing unit

TOOLS
Gauge manifold, charging hoses, service wrench, 10-in adjustable wrench, pliers, Phillips screwdriver, straight screwdriver, 1/4-in socket set

SUPPLIES
Shop towel

JOB 2–1
Installing service gauges and recording high- and low-side pressure readings

PROCEDURE
1. Connect flexible charging hoses to gauge manifold. (Depressor pin is in the end of the hose, opposite the gauge manifold.)
2. Turn gauge manifold valves clockwise to their "off" position.
3. Turn off the unit if it is running.
4. Locate the suction line and install the low-side gauge hose on the service valve.
5. Purge air from the hose using the low-side manifold valve.
6. Locate the high-side service valve and connect the high-side manifold gauge line.
7. Purge air from the high-side hose through the center manifold port.
8. If the compressor has double-seated suction and discharge service valves, turn the valve stem counterclockwise to close the gauge port when installing a charging hose.
9. Turn on the unit and check the operating pressures.

■ JOB SHEET 2–2 ■

Name _____

Score _____ Date _____

PERFORMANCE OBJECTIVE

Given a dial-stem pocket thermometer, electronic temperature tester, an operating AC unit with gauges installed, and a condensed pressure-temperature chart, within 1 hour measure the intensity of heat. Then, identify the three methods of heat transfer at the outdoor and indoor sections of an operating central AC system.

REFERENCE

Heating, Ventilating, and Air Conditioning Fundamentals (Chapter 2), pressure-temperature chart (Figure A–1)

EQUIPMENT

Central AC system

TOOLS

Temperature-testing equipment, hand tools

SUPPLIES

Duct tape

JOB 2–2

Identifying methods of heat transfer and recording intensity of heat

PROCEDURE

1. Connect three test leads from the electronic thermometer to the suction line, discharge line, and liquid line.
2. Compare temperature readings with pressure-gauge readings. Do temperatures match those indicated on condensed pressure-temperature chart?
3. Which one does not?
4. Underline the correct answer. It is higher than gauge reading because of *(a)* latent heat, *(b)* superheat, or *(c)* specific heat.
5. The method of heat transfer is by _____.
6. Measure air temperature as close to crankcase heater as possible without touching it with the temperature sensor. The method of heat transfer is by _____.
7. Measure the air temperature upstream and downstream of the evaporator coil. The method of heat transfer is by _____.

■ MULTIPLE-CHOICE TEST ■

Name _____

Score _____ Date _____

DIRECTIONS

Circle the letter that best answers the following multiple-choice questions.

1. Heat is a form of:

   a. work   c. matter
   b. power   d. energy

2. Power may be defined as the:

   a. energy expended
   b. horsepower required
   c. rate of doing work
   d. applied energy

3. The molecular theory of heat states that the molecules of any substance are disturbed when:

   a. power is used
   b. energy is expended
   c. work is performed
   d. heat is applied

4. The heat intensity of a substance can be measured and expressed in:

   a. degrees   c. molecular motion
   b. energy   d. change of state

5. A British thermal unit (Btu) of heat is the amount of heat required to be added to one pound of water to change its temperature:

   a. to boiling point   c. 1°C
   b. 1°F   d. 32°F

6. The specific heat of water is:

   a. 0.5   c. 1.0
   b. 0.24   d. 1.5

7. Quantity of heat is expressed in

   a. degrees   c. intensity
   b. Btu   d. horsepower

8. 34°C is:

   a. 93.8°F   c. 93.4°F
   b. 93.6°F   d. 93.2°F

9. Sensible heat is the heat added to a substance to:

   a. change its state
   b. raise its temperature
   c. cause evaporation
   d. freeze it

10. Latent heat of fusion is heat absorbed by a solid during conversion to:

    a. vapor   c. solid
    b. liquid   d. ice

11. Total heat is:

    a. absolute zero
    b. the sum of sensible and latent heat
    c. the heat that causes a change of state
    d. the heat of evaporation

12. The number of Btu removed in 24 h equaling 1 ton of refrigeration effect is:

    a. 2000 Btu   c. 33,000 Btu
    b. 12,000 Btu   d. 288,000 Btu

13. The specific heat of 35 lb of substance that requires 242 Btu to increase its temperature from 62 to 72°F is:

    a. 0.6149   c. 0.6491
    b. 0.6419   d. 0.6914

14. Heat transferred by conduction from one body to another is done:

    a. via liquids   c. via solids
    b. via heat rays   d. via radiation

15. Radiated heat is reflected by a:

    a. clean surface
    b. smooth, light surface
    c. dark surface
    d. rough surface

■ WRITTEN EVALUATION ■

1. Name the high and low evaporator refrigerant boiling temperatures for an air-conditioning application.
2. Name three malfunctions that would cause the refrigerant in the evaporator to boil below the normal temperature range.
3. How do you determine the approximate high-side pressure?

Unit 3
COMPILING PERFORMANCE DATA

■ INTRODUCTION

In order to diagnose the cause of a hermetic compressor failure, we must first know what conditions exist when the system is working properly. This information can be called *prerecorded findings*. The service technician can then compare the results of his or her mechanical and electrical tests so that the problem can be isolated.

The tasks in Unit 3 are designed to help you organize a systematic approach for testing and replacing hermetic compressors.

■ PRESENTATION

Located below is the standard, or prerecorded, data that the service technician should always consult. These are readily available from past experience and references:

Past experience
Manufacturer's service data
Manufacturer's electrical guidebook
Compressor serial plate information
Unit information tag
Manufacturer's service literature

Equipment supply houses generally have on hand the service literature required to service their equipment. This literature is often free or sold at a minimal charge.

This literature is also very important. Hermetic compressors, for example, usually do not have an oil sight glass. So, you'll need the manufacturer's literature to determine the correct oil charge. This information is compiled by the field service mechanic and should be carried in the service truck.

Motor current ratings, also, are often difficult to obtain. The nameplate may show only the locked rotor amps (LRA), which is the stalled or starting current draw of the motor. (See Table 3–1 for motor current ratings.)

Table 3-1   Motor Current Ratings

hp	Single-Phase, A-C Motors[1]						Three-phase, Induction-Type, A-C Motors[2]							
	120 V		240 V		440 V		120 V		240 V		440 V		550 V	
	Full Load	Locked Rotor	Full Load	Locked Rotor	Full Load	Locked Rotor	Full Load	Locked Rotor	Full Load	Locked Rotor	Full Load	Locked Rotor	Full Load	Locked Rotor
1/6	3.2	19.2	1.6	9.6										
1/4	4.6	27.6	2.3	13.8										
1/3	5.5	33.0	2.8	16.8	...	...								
1/2	7.4	44.4	3.7	22.2	...	...	4.0	24.0	2.0	12.0	1.0	6.0	.8	4.8
3/4	10.2	61.2	5.1	30.6	...	...	5.6	33.6	2.8	16.8	1.4	8.4	1.1	6.6
1	13.0	78.0	6.5	39.0	...	...	7.0	42.0	3.5	21.0	1.8	10.8	1.4	8.4
1 1/2	18.4	110.4	9.2	55.2	...	...	10.0	60.0	5.0	30.0	2.5	15.0	2.0	12.0
2	24.0	144.0	12.0	72.0	...	...	13.0	78.0	6.5	39.0	3.3	19.8	2.6	15.6
3	34.0	204.0	17.0	102.0	...	...	...	...	9.0	54.0	4.5	27.0	4.0	24.0
5	56.0	336.0	28.0	168.0	...	...	...	...	15.0	90.0	7.5	45.0	6.0	36.0
7 1/2	80.0	480.0	40.0	240.0	21.0	126.0	...	...	22.0	132.0	11.0	66.0	9.0	54.0

[1]For full-load currents of 208- and 200-V motors, increase the corresponding 240-V, motor full-load current by 10 and 15 percent, respectively. To find the locked-rotor currents for the 208- and 200-V motors, multiply the full-load currents of the motors by 6.

[2]For full-load currents of 208- and 200-V motors, increase the corresponding 240-V motor full-load current by 6 and 10 percent, respectively. To find the locked-rotor currents for the 208-V and 200-V motors, multiply the full-load currents by 6.

Courtesy Honeywell Corporation.

■ JOB SHEET 3-1 ■

Name _____

Score _____ Date _____

PERFORMANCE OBJECTIVE
Given a running condensing unit, gauge manifold and hoses, volt-ohm-milliameter, amp meter, and necessary tools, check the compressor operation. Your findings must agree with the prerecorded findings.

REFERENCE
Heating, Ventilating, and Air Conditioning Fundamentals (Chapter 3); *Hermetic Compressor Service Data,* and *Electrical Service Parts Guide Book,* Tecumseh Products Company

EQUIPMENT
Operable air-conditioning unit

TOOLS
VOM meter, amp meter, gauge manifold and hoses, hand tools

SUPPLIES
Shop towel, notebook

JOB 3-1
Checking compressor operation

PROCEDURE
1. Install gauges and start unit.
2. Check unit starting and running pressures.
3. Check amps at unit starting and running.
4. Check pressures to determine if unit is pumping properly.
5. Check motor mounts for proper tension.
6. Check for vibration and rattle.
7. Check system for proper amount of oil.
8. Check system for refrigerant overcharge.
9. Remove gauges from unit.
10. Compare your findings with the prerecorded findings.

■ JOB SHEET 3-2 ■

Name _____

Score _____/_____ Date _____

PERFORMANCE OBJECTIVE
Given a hermetic compressor in the system, service valves, gauge manifold and hoses, volt-ohm-milliameter, and amp meter supplies and tools, test the compressor electrically and mechanically recording pressure, electrical ratings, and oil-test results. Your recorded data must match prerecorded data.

REFERENCE
Heating, Ventilating, and Air Conditioning Fundamentals (Chapter 3); *Installation and Service Application Manual*, Tecumseh Products Company; *Getting to Know Your Electrical Test Instrument*, A. W. Sperry Company

EQUIPMENT
Operable air-conditioning unit with hermetic compressor

TOOLS
Gauge manifold and hoses, pinch-off tool, tube cutter, air-acetylene torch kit, service wrench, striker, pliers, swage tool and hammer, VOM meter, amp meter

SUPPLIES
Shop towel, oil-test kit, Sil-Fos and flux, sandcloth

JOB 3-2
Testing hermetic compressor and recording pressure, electrical ratings, and oil-sample test results

PROCEDURE
1. Install gauges.
2. Check pressure.
3. Pinch suction line near compressor.
4. Pump compressor down to 26 in of vacuum.
5. Reopen suction line.
6. Remove compressor terminal cover.
7. Check amperes with amp meter while compressor is running.
8. Disconnect power supply.
9. Remove start, run, and common wires from terminals.
10. Check internal wires for continuity and ground.
11. Check overload and relays for continuity.
12. Complete compressor change-out form.
13. Test oil sample with acid test kit (disconnect compressor if needed).
14. Reinstall compressor to original condition.
15. Record data and compare to prerecorded data.

■ MULTIPLE-CHOICE TEST ■

Name _____

Score _____ Date _____

DIRECTIONS
Circle the letter that best answers the following multiple-choice questions.

1. The highest resistance found will be between compressor motor terminals:

   a. R–S
   b. C–R
   c. S–C

2. Hermetic compressors from 1 to 5 hp will most likely use a:

   a. hot-wire thermal relay
   b. current relay with a start capacitor
   c. resistance-start, induction-run motor
   d. potential relay

3. The start winding:

   a. has a higher resistance than the common winding
   b. has a lower resistance than the run winding
   c. has a higher resistance than the run winding

4. The coil on a potential relay:

   a. is wired in series with common
   b. is wired in parallel with the start winding
   c. is wired in series with the run winding
   d. has a low impedance to current flow

5. The coil on a current relay:

   a. has only a few turns of wire
   b. is wired in parallel with the run winding
   c. is wired in series with the start winding

6. To increase the torque of a PSC compressor motor:

   a. install a hard-start kit
   b. increase the size of the start capacitor
   c. use a higher microfarad run capacitor
   d. install a run capacitor with a higher voltage rating

7. Starting torque will increase progressively with the following motors:

   a. CSR, RSIR, PSC
   b. RSCR, CSIR, split phase
   c. current, potential, hermetic
   d. PSC, CSIR, CSR

8. Motor current should not exceed:

   a. LRA
   b. WPA
   c. FLA
   d. PSC

9. A stuck compressor can be indicated by:

   a. locked rotor amps
   b. full-load amps
   c. 26-in vacuum
   d. low resistance to ground reading

10. Refrigeration oil should have:

    a. low pour point, low wax content, low carbon-forming tendencies, low floc point
    b. low aromatics, high APl gravity, lower specific gravity, high floc point
    c. high wax content, high pour point, high viscosity
    d. 10W-40 viscosity

■ WRITTEN EVALUATION ■

1. Describe the procedures for a mild burnout compressor replacement.
2. Outline the steps for a severe burnout compressor replacement.
3. How do you perform a mechanical efficiency test on a hermetic compressor?

Unit 4

SERVICING EVAPORATORS AND CONDENSERS

■ INTRODUCTION

The following tasks can prepare you for an entry-level job with a company specializing in contract maintenance of water-cooled condensing unit accessories. Before completing the tasks:

1. Read through the procedure steps to determine what is to be done.

2. Check your assigned AC unit to see if the task matches the type of heat-exchange equipment on your unit.
3. Reread the material pertaining to the assigned task.
4. Plan a field trip if the equipment to perform the following tasks is not readily available.

■ JOB SHEET 4–1 ■

Name _____

Score _____ Date _____

PERFORMANCE OBJECTIVE
Given a residential unit with a malfunctioning air-cooled condenser
and necessary equipment and supplies, check and service an air-
cooled condenser and restore the unit to its proper operation.

REFERENCE
Heating, Ventilating, and Air Conditioning Fundamentals (Chapter 4); material on air-cooled condensers

EQUIPMENT
Air-cooled condensing unit, or air-cooled package unit

TOOLS
Fin-comb, manifold gauges and hoses, hand tools, thermometer,
pressure-temperature chart (Figure A-1)

SUPPLIES
Tank of CO_2

JOB 4–1
Checking and servicing an air-cooled condenser

PROCEDURE
1. Disconnect power.
2. Gain access to condenser; remove panels if necessary.
3. Check condenser for bent fins; straighten if necessary.
4. Check condenser for any matter restricting airflow. Clean
    with a brush or blow out dirt with CO_2 if necessary.
5. Check for worn bearing on fan (the shaft should not have
    any up or down play). Replace or lubricate fan if neces-
    sary.
6. Install gauge manifold and hoses.
7. Replace panels if removed to prevent fan from bypassing
    condenser air.
8. Restore power.
9. Turn on unit and observe for proper operation.
10. Record readings:

    Head pressure: _____
    Suction pressure: _____
    Ambient temperature: _____
    Compressor FLA: _____
    Actual running amps: _____

11. Disconnect gauges and turn off power.

■ JOB SHEET 4–2 ■

PERFORMANCE OBJECTIVE

Given a unit with a finned evaporator and blower assembly, and necessary tools, visually check the evaporator for airflow obstructions, or plugged condensate drain. Findings must match prerecorded findings.

REFERENCE

Heating, Ventilating, and Air Conditioning Fundamentals (Chapter 4)

EQUIPMENT

Unit with a DX blower coil

TOOLS

Flashlight, screwdrivers, hand tools

SUPPLIES

Shop towel, tank of CO_2 and hose

JOB 4–2

Checking a direct-expansion finned evaporator

PROCEDURE

1. Disconnect power to the unit.
2. Visually inspect evaporator for obstructions or kinked tubing.
3. Shine light through fins and look for any obstructions.
4. Blow CO_2 through drain-pan condensate-line connection. (Seal opening around air line with shop towel.)
5. Record your findings.

■ JOB SHEET 4–3 ■

Name _____

Score _____ Date _____

PERFORMANCE OBJECTIVE
Given a water tower and the necessary tools and chemicals, service the water tower. The serviced water tower must operate without noise or vibration.

REFERENCE
Heating, Ventilating, and Air Conditioning Fundamentals (Chapter 4); information on cooling towers and evaporative condensers

EQUIPMENT
Operable unit with a cooling tower or evaporative condenser

TOOLS
Two pipe wrenches, 12-in adjustable wrench, hand tools, sling psychrometer, set of Allen screw wrenches

SUPPLIES
Recommended chemicals for water treatment

JOB 4–3
Servicing and treating a water tower

PROCEDURE
1. Check spray nozzles while tower is in operation.
2. Disconnect power.
3. Check pump inlet screen; clean if needed.
4. Check for algae or scale in tower; clean spray nozzles if needed. Drain and flush basin if necessary.
5. Check belts for proper tension and wear.
6. Check float valve for leaks and proper water level.
7. Check pulleys for proper wear and alignment.
8. Check bearings for noise and wear. Lubricate bearings.
9. Treat water tower with chemicals according to manufacturer's specifications.
10. Restore power and watch for any noise or vibration.

■ JOB SHEET 4–4 ■

PERFORMANCE OBJECTIVE
Given a unit with a water-cooled condenser with removable end plates, electric drill, nylon brush (wire brush if nylon not available), goggles, and necessary tools, service the condenser. When servicing is completed, the water flow and drain should be free of obstructions.

REFERENCE
Heating, Ventilating, and Air Conditioning Fundamentals (Chapter 4)

EQUIPMENT
Unit with a shell and tube water-cooled condenser

TOOLS
½-in electric drill, extension cord, rod and brush, socket set, hammer

SUPPLIES
End-plate gaskets

JOB 4–4
Servicing a water-cooled condenser

PROCEDURE
1. Turn off the system.
2. Turn off the water supply and drain condenser.
3. Remove condenser end plates.
4. Assemble drill rod and brush.
5. Run brush through the tubes.
6. Flush the tubes with water.
7. Inspect the end-plate gaskets; replace if necessary.
8. Remount the end plates.
9. Turn the water on.
10. Check for leaks.
11. Check the pump strainers.
12. Check the head pressure heat dissipation.

■ FILL IN THE BLANKS ■

Name _____

Score _____ Date _____

DIRECTIONS

Fill in the blanks with the correct missing words.

1. The condenser first removes _____ heat, and then removes _____ heat.
2. Three condensing mediums used in refrigeration are: _____, _____, and _____.
3. Two types of commonly used condenser fans are: _____ and _____.
4. The water flows _____ the tubes, and the refrigerant flows in a counterflow direction _____ the tubes in a shell and tube condenser.
5. There are three basic types of condensers: _____, _____, _____ cooled.
6. There are two basic types of evaporators: the _____ and _____ types.
7. The proper selection of an evaporator depends on the _____.
8. In a direct expansion chiller the _____ flows inside the tubes.
9. In a fully charged cooler the refrigerant _____ the tubes.
10. Secondary refrigerants are usually _____ or _____.

■ WRITTEN EVALUATION ■

1. The evaporator design temperature is 40°F (4.4°C). What would you look for if you found the evaporator iced up?
2. How do you determine if there is air in the system?
3. List the reasons for wanting to change from water-cooled units to air-cooled units.

Unit 5

METERING DEVICES

■ INTRODUCTION

The apprentice often sees the metering device as either a complex device or a cure-all for a malfunctioning system. Consequently, an apprentice's misguided approach could result in another unnecessary visit to correct the original problem. The technician would have to readjust the thermostatic expansion valve's superheat, and may even need to bleed off an overcharge of refrigerant.

The following tasks should erase any doubts about what a metering valve is supposed to do: namely, to keep the evaporator supplied with enough refrigerant to satisfy the load condition without allowing liquid refrigerant to enter the suction line.

Use Table 5–1 when completing the first performance task.

TABLE 5–1   Capillary tube selection R-22 (high temperature).

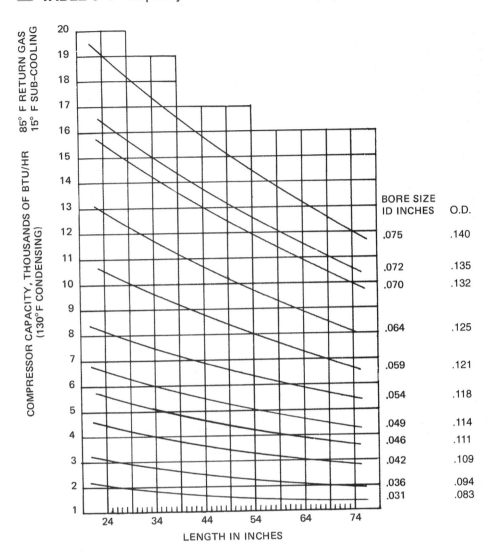

BORE SIZE ID INCHES	O.D.
.075	.140
.072	.135
.070	.132
.064	.125
.059	.121
.054	.118
.049	.114
.046	.111
.042	.109
.036	.094
.031	.083

■ JOB SHEET 5–1 ■

Name _____

Score _____ Date _____

PERFORMANCE OBJECTIVE
Given an operating air-conditioning unit with a capillary-tube metering device, gauges, electronic temperature tester, and necessary hand tools, determine within a 1/2-hour running period if the cap tube is properly sized.

REFERENCES
Heating, Ventilating, and Air-Conditioning Fundamentals (Chapter 5)

EQUIPMENT
Residential AC system with a capillary tube

TOOLS
Electronic temperature test equipment, gauges, hand tools

SUPPLIES
Roll of insulating tape

JOB 5-1
Working with capillary-tube system

PROCEDURE
1. Install the gauges.
2. Start the unit and record the suction reading _____ psi. Text requirements: _____ minimum; _____ maximum.
3. Connect temperature test lead No. 1 to the return ell near the center of evaporator. Connect the No. 2 lead to the bottom of suction line, 6 in from the compressor. Record the temperature difference: _____. Text requirements: _____ minimum; _____ maximum.
4. Connect the No. 3 lead to the evaporator outlet and record the temperature difference between lead No. 1 and lead No. 2. On low demand this temperature difference should be: (check one) the same _____, lower _____, higher _____. On peak demand this temperature difference should be: the same _____, lower _____, higher _____.
5. Measure the cap-tube outside diameter (OD) with a micrometer and determine the Btu/h capacities from the inside diameter (ID) measurements (see Table 5-1). OD: _____; ID: _____; Btu/h: _____.

JOB SHEET 5-2

PERFORMANCE OBJECTIVES
Given a refrigeration system with a restricted automatic expansion valve, gauges, and necessary hand tools, clean or replace the valve inlet screen and adjust the valve to constant pressure maintaining a 32 to 40°F (0° to 4.4°C) coil outlet temperature (slightly higher at peak demand).

CRITERION-REFERENCED MEASURE
Remove restriction and adjust pressure of automatic expansion valve.

REFERENCE
Heating, Ventilating, and Air Conditioning Fundamentals (Chapter 5)

EQUIPMENT
Training unit with automatic expansion valve

TOOLS
Gauge manifold and necessary hand tools

JOB 5-2
Adjusting an automatic expansion valve (AXV)

PROCEDURE
1. Install gauges.
2. Start the unit and record suction pressure. The pressure should correspond to a temperature between 0 and 5°C.
3. Observe the liquid line at the valve inlet. A restricted valve is often indicated by condensation forming on the liquid line and a partially frosted evaporator.
4. Remove pressure from the valve.
5. Turn the system off.
6. Remove the strainer and clean.
7. Evacuate the system that was opened for strainer removal.
8. Put the system back in operation.
9. Check the suction pressure. The pressure should be constant.
10. Adjust the valve to zero or a slightly higher superheat at the coil outlet.
11. Check the suction line temperature at the compressor inlet. The temperature must be above the coil outlet temperature, but not above 18°C.

■ JOB SHEET 5–3 ■

Name _____

Score _____ Date _____

PERFORMANCE OBJECTIVE
Given an operating air-conditioning system with an externally equalized thermostatic expansion valve (TXV), gauge manifold, temperature tester, pressure-temperature chart, and necessary hand tools, adjust the valve to 10° superheat.

CRITERION-REFERENCED MEASURE
Check and adjust the TXV to a 10° superheat setting.

REFERENCE
Heating, Ventilating, and Air Conditioning Fundamentals (Chapter 5); "What's Your Superheat?" Sporlan Service Literature

EQUIPMENT
Central station air-conditioning unit with an externally equalized TXV

TOOLS
Gauge manifold, electronic temperature tester, hand tools

SUPPLIES
Duct tape or insulating tape

JOB 5-3
Adjusting thermostatic expansion valve (TXV) superheat

PROCEDURE
1. Check the power assembly. Be sure excess tubing is neatly coiled and see that there are no rubbing or kinked areas.
2. Check the bulb for correct location and tight connection.
3. Install the gauge manifold.
4. Attach the thermometer to the suction line with a bulb clamp, or secure with duct tape.
5. Find the evaporator saturation (boiling) temperature by adding 2 psi to the suction pressure at the compressor, or by converting the evaporator outlet pressure directly to temperature with the pressure-temperature chart.
6. If the coil outlet vapor temperature and the evaporator saturation temperature are not 10°F apart, adjust the valve.
7. Remove the gauges and thermometer.

■ JOB SHEET 5–4 ■

Name _____
Score _____ Date _____

PERFORMANCE OBJECTIVES
Given a heat pump with a thermal-electric expansion valve, potentiometer, electric thermometer, volt-ohm-milliameter, (1) modulate
the valve from its normal 0° superheat control at sensor location,
and (2) discover how a thermistor changes its resistance value with
a temperature change.

CRITERION-REFERENCED MEASURE
Given a heat pump with a thermal electric expansion valve, modulate the valve with a bidirectional flow.

REFERENCE
Heating, Ventilating, and Air Conditioning Fundamentals (Chapter 5); heat-pump application

EQUIPMENT
Heat pump with thermal-electric expansion valve, two sensors that
connect to the outdoor coil and indoor coil, and four-way valve piping

TOOLS
Electronic temperature tester, volt-ohm-milliameter, watt meter

SUPPLIES
Wire nuts, 100-Ω potentiometer, two test leads (No. 16 AWG) with
alligator clips

JOB 5-4
Testing a thermal-electric expansion valve and thermistor sensors

PROCEDURE
1. With power off, measure the resistance of the valve motor.
_____ Ω (Singer Model 625) or other _____.
2. Measure the resistance of the thermistor. _____ Ω Typical
readings at 77°F (25°C) for three (Model 625) sensors 19,
50, and 100 Ω.
3. Reconnect the wiring with sensors in series with the valve
motor and low-voltage power source (24 V).
4. Set room thermostat for a call for cooling and operate unit.
5. Connect voltmeter across thermistor and measure voltage
drop, if any.
Indoor coil sensor: _____ volts
Outdoor coil sensor: _____ volts
6. Connect the temperature test leads to indoor and outdoor
refrigerant lines near thermistor sensors.
7. Turn off the unit and disconnect the power source.

8. Record thermistor resistance readings
Indoor coil: _____; temperature: _____ Ω
Outdoor coil: _____; temperature: _____ Ω
9. From the difference in ohm readings (steps 2 and 8) determine the following:

    a. An increase in temperature: _____ resistance
    b. A drop in temperature: _____ resistance

10. Disconnect one lead of the valve motor and connect the potentiometer in series with the valve motor. Use test leads with alligator clips. By using the center tap and either outside connection of the potentiometer it serves as a rheostat (variable resistor). (See Figure 5–1.)
11. With the rheostat connected in the series circuit, operate the unit in the cooling cycle and record the following. Temperature at outdoor coil sensor with rheostat turned
Fully clockwise: _____°F; _____°C
Midpositioned: _____°F; _____°C
Fully counterclockwise: _____°F; _____°C
Temperature at indoor coil sensor with rheostat turned
Fully counterclockwise: _____°F; _____°C
Midpositioned: _____°F; _____°C
Fully counterclockwise: _____°F; _____°C
12. Turn off unit and remove test equipment.

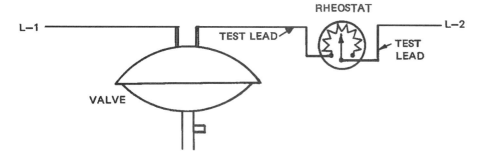

FIGURE 5–1   Connecting the potentiometer.

■ MULTIPLE-CHOICE TEST ■

Name _____

Score _____ Date _____

DIRECTIONS
Circle the letter that best answers the following
multiple-choice questions.

1. An orifice plate metering device is generally found on:

   a. centrifugal units.
   b. window-type air conditioners
   c. heat pumps
   d. lithium bromide units
   e. air-cooled package units

2. Capillary tube sizes are governed by:

   a. length and width
   b. length and inside diameter
   c. length and outside diameter
   d. outside diameter and application
   e. refrigerant and outside diameter

3. If an existing capillary tube is shortened
   and the unit is fully charged:

   a. the operating suction pressure will
      rise
   b. the operating suction pressure will
      drop
   c. the coil superheat will rise
   d. the coil superheat will drop

4. With an automatic expansion valve:

   a. a decrease in spring pressure opens
      the valve
   b. an increase in spring pressure raises
      the evaporator's superheat
   c. atmospheric pressure closes the valve
   d. atmospheric pressure opens the valve

5. With an automatic expansion valve:

   a. a sudden increase in load will open the
      valve
   b. a sudden increase in load will close
      the valve
   c. a drop in suction pressure will open
      the valve

6. An automatic expansion valve without a
   bleed orifice cannot be used on the following compressor motor:

   a. shaded pole
   b. unshaded pole
   c. rapid balancing pole
   d. permanent split capacitor

7. The following type thermostatic expansion valve is required when the evaporator has a large pressure drop:

   a. rapid pressure balancing
   b. internally equalized
   c. externally equalized
   d. bleed type

8. The power assembly on a TXV refers to:

   a. diaphragm, capillary tube, and bulb
   b. capillary tube, bulb, and superheat
      spring
   c. diaphragm, capillary tube, bulb, and
      superheat spring
   d. type of bulb charge

9. The closing force with a TXV is:

   a. power assembly
   b. power assembly plus superheat spring
   c. evaporator pressure plus spring pres-
      sure

10. An increase of voltage supplied to a ther-
    mal electric valve will:

    a. open the valve
    b. close the valve

■ WRITTEN EVALUATION ■

1. Describe how to check superheat on an externally equalized TXV.
2. Explain how to adjust an automatic expansion valve.
3. Describe how to test a capillary tube system for a proper charge without the use of a gauge manifold.

Unit 6

PRESSURE-ENTHALPY DIAGRAMS

■ INTRODUCTION

The pressure-enthalpy diagram, also called the Mollier diagram, charts a refrigerant's pressure, heat, and temperature properties. And with the help of the R-12 pressure-enthalpy diagram, you can determine the quantity of heat that is picked up in the evaporator for each pound of R-12 that is circulated. By adding the heat energy required to circulate the refrigerant (entropy), you can also determine the condenser load. Moreover, if the heat load is known, the constant volume determined from the plotted cycle can be used to find the required compressor displacement (see Figure 6–1).

The next step is to apply the Mollier diagram to an actual air-conditioning application. For this, refer to Figure 6–2, which charts R-22 in an AC unit.

■ PRESENTATION

The refrigeration cycle starts at the orifice of the metering device. To determine the cycle starting point (intersection of the condensation line and the expansion line), proceed as follows:

1. Add 14.7 psi to the high-side gauge reading.
2. Measure the temperature of the liquid line at the metering device inlet.
3. Refer to the pressure-temperature chart for the corresponding liquid line temperature-pressure relationship (see Figure A-1).
4. If the pressure in step 3 corresponds to the pressure in step 1, the starting point will be located on the saturated liquid line.
5. If the pressure is lower in step 3 than in step 1, the starting point will be found in the subcooled zone on the constant-pressure

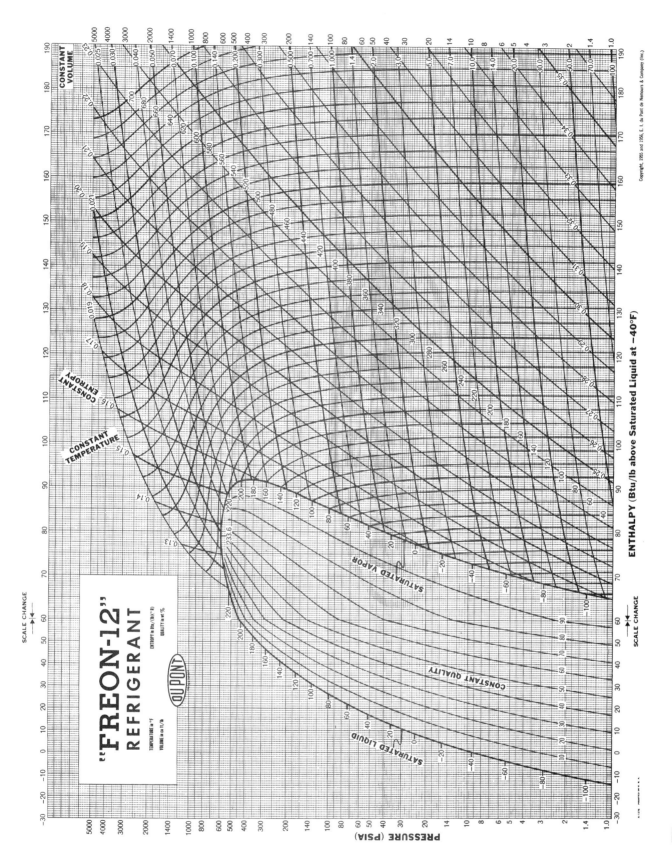

FIGURE 6–1 Freon-12 pressure-enthalpy diagram. (*E.I. DuPont de Nemours & Co., Inc.*)

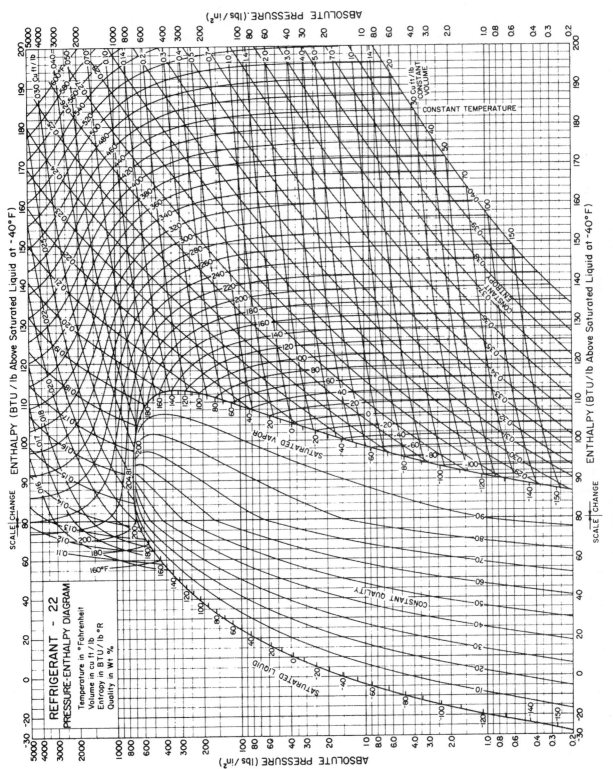

© 1964, E. I. du Pont de Nemours & Company, Inc. (Used by permission.)

FIGURE 6–2 Freon-22 pressure-enthalpy diagram.

condensation line that corresponds to the pressure in step 1, and at a point that matches the temperature of the subcooled (below saturation) liquid line.

Moreover, note the constant-quality lines that are found on the R-22 diagram in the mixed zone. The *constant quality* is the percentage of liquid (flash gas) that is instantly boiled off due to expansion. This cools the remaining liquid to the evaporator saturation (boiling) temperature. The more flash gas, the lower the amount of liquid left to boil in the evaporator. Consequently this yields a lower net refrigeration effect.

Use the diagram as a reference for matching specific locations of a refrigeration system plotted on a pressure-enthalpy diagram.

■ JOB SHEET 6–1 ■

Name _____

Score _____ Date _____

PERFORMANCE OBJECTIVE
Given a Freon-22 pressure-enthalpy diagram, draw a skeletal diagram in your notebook and insert a letter to identify the basic processes of a vapor compression cycle such as shown in Figure 6–2 with a maximum of one error.

REFERENCE
Heating, Ventilating, and Air Conditioning Fundamentals (Chapter 6); R-22 pressure-enthalpy diagram

EQUIPMENT
Air-conditioning unit

TOOLS
Hand tools

SUPPLIES
Notebook, pressure-enthalpy diagram, ruler

JOB 6–1
Drawing a skeletal pressure-enthalpy diagram and identifying the basic processes of a vapor compression cycle

PROCEDURE
1. On a sheet of notebook paper, draw a saturated liquid and vapor curve line similar to that shown in Figure 6–2.
2. Draw a condensation line and place the correct letter _____ where the condensation line intersects the saturated-liquid line at 100°F (see Figure 6–2).
3. Draw the evaporation line and indicate 40°F at the saturated-liquid-line intersection (see Figure 6–2).
4. Draw the expansion line and indicate the correct letter _____ where the expansion line and evaporation line intersect (see Figure 6–2).
5. The constant quality indicated at step 4 is approximately _____ percent.
6. At point _____ the refrigerant in the evaporator is in a saturated-vapor state (see Figure 6–3).
7. The TXV power assembly bulb falls on the constant pressure evaporation line at point _____ 50°F superheated-vapor zone (see Figure 6–3).
8. The compressor inlet _____ is located 2°F below the evaporator outlet point.
9. The compressor discharge point _____ represents the entropy line that intersects with the constant-pressure condensation line (Figure 6–3).

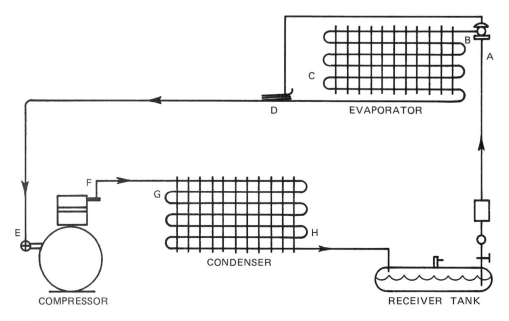

FIGURE 6–3   Refrigeration cycle.

10. The first pass in the condenser releases superheat. There-
    fore, point _____ is located at the intersection of the con-
    densation- and saturated-vapor lines.
11. At times the condenser may cool the liquid lower than
    point _____, which represents saturated liquid. This
    means that the refrigerant leaving the condenser would fall
    in the _____ zone and _____ gas would be reduced.

■ JOB SHEET 6–2 ■

Name _____

Score _____ Date _____

PERFORMANCE OBJECTIVE
Given an operating AC unit, pressure-enthalpy diagram, gauge manifold, and temperature-test equipment, sketch a pressure-enthalpy cycle, record actual temperatures and pressures, and insert on the diagram the heat content per pound and specific volume per pound.

REFERENCE
Heating, Ventilating, and Air Conditioning Fundamentals (Chapter 6); pressure-enthalpy diagram; refrigerant data from E. I. DuPont de Nemours & Co., Inc.

EQUIPMENT
Operable AC unit

TOOLS
Electronic temperature tester, hand tools, gauge manifold and hoses

SUPPLIES
Notebook, pressure-enthalpy diagram, ruler, duct tape (Figure 6–4)

JOB 6–2
Sketching a pressure-enthalpy diagram and indicating proper values for the basic processes of a vapor-compression cycle

PROCEDURE
1. Sketch a pressure-enthalpy diagram and identify the lines of the vapor compression cycle.
2. Turn on your AC unit and record on your diagram the following temperatures:
    Saturated liquid high-side: _____°F; _____°C
    Saturated liquid low-side: _____°F; _____°C
    Evaporator outlet superheat degrees: _____°F; _____°C
    Discharge temperature: _____°F; _____°C
    Condensing temperature: _____°F; _____°C
    Condenser subcooling: _____°F; _____°C
    Temperature drop from condenser outlet to metering device:_____°F; _____°C
    Suction-line temperature rise: _____°F; _____°C
3. The constant quality approximated from our diagram indicates _____percent of the liquid entering the evaporator was lost due to flash gas.
4. The net refrigeration effect lost due to flash gas is the enthalpy difference between the low-side saturated-vapor line and the expansion line, or _____Btu/lb.

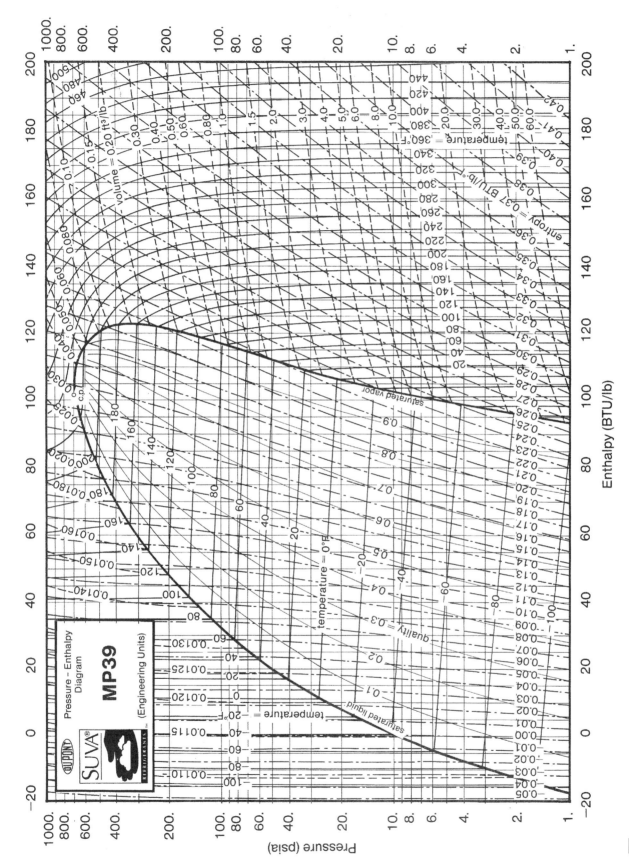

FIGURE 6-4 SUVA® MP39 pressure vs. enthalpy.

5. The net refrigeration effect is the enthalpy difference be-tween the expansion line and the low-side saturated-vapor line, or _____Btu/lb.
6. From the low-side saturated-vapor line of the cycle you have drawn, follow on a parallel plane a constant volume line. The constant volume is _____ft/lb.
7. Measure the temperature of the discharge line: _____°F; _____°C
8. Superimpose your cycle over an actual pressure-enthalpy diagram. Does the temperature you recorded (step 7) match the diagram? _____
9. The entropy line represents the inlet and outlet of the com-pressor. Our cycle indicates _____Btu/lb entropy value.
10. The temperature at the suction line and 6 in upstream of the compressor is: _____°F; _____°C. This temperature should not exceed 65°F (18.3°C).
11. The temperature difference between step 7 and step 10 in-dicates that the refrigerant leaving the compressor is super-heated: _____°F; _____°C.

■ JOB SHEET 6–3 ■

Name _____

Score _____ Date _____

PERFORMANCE OBJECTIVE
Given an SI metric pressure-enthalpy diagram and a sheet of vellum paper, draw a skeletal diagram and record the metric measurements for an R-22 system operating at 196 psig head pressure and 58 psig suction pressure.

REFERENCE
Heating, Ventilating, and Air Conditioning Fundamentals (Chapter 6); SI metric pressure-enthalpy diagram (available from Carrier Corp., Allied Chemical Corp., and E. I. DuPont de Nemours & Co., Inc.)

TOOLS
Drafting tools

SUPPLIES
SI metric pressure-enthalpy diagram, vellum paper

JOB 6–3
Drawing a skeletal diagram and recording metric measurements

PROCEDURE
1. Place the sheet of vellum over the SI metric pressure-enthalpy diagram. Trace the saturated liquid and vapor lines.
2. Convert 196 psig head pressure to _____psia.
3. Multiply psia (step 2) × 6.895 _____kPa.
4. Locate step 3 (Pa) on the diagram and draw the head pressure line across the chart.
5. Convert 58 psig suction pressure to _____psia.
6. Multiple psia (step 5) × 6.895 _____kPa.
7. Locate step 6 (Pa) and draw the suction pressure line across the chart.
8. Draw the expansion line from the saturated liquid (step 3). The enthalpy is _____kJ/kg.
9. The flash gas is _____kJ/kg.
10. The constant volume at the saturated-vapor suction pressure is _____m^3/kg.
11. The constant entropy value from the saturated-vapor line is _____kJ/kg.
12. The net refrigeration effect (NRE) is _____kJ/kg.
13. The kJ/kg condenser load is _____kJ/kg.
14. The compressor discharge temperature is _____°C.

■ JOB SHEET 6–4 ■

Name _____

Score _____ Date _____

PERFORMANCE OBJECTIVE
Given an operating unit with R-12 refrigerant, complete System Data Sheet (Figure 6–5). Convert system to MP 39 refrigerant. Performance should compare to R-12 data.

REFERENCE
Heating, Ventilating, and Air Conditioning Fundamentals (Chapter 6); E. I. DuPont de Nemours & Co. Data Sheet and retrofit checklist for MP 39

TOOLS
Gauge manifold, sling psychrometer, electronic temperature tester, amprobe, hand tools, leak detector, vacuum pump, recovery unit, oil pump

SUPPLIES
XH9 desiccant drier, alkylbenzene oil, MP 39 refrigerant

JOB 6–4
Retrofit R-12 unit to MP 39

PROCEDURE
1. Complete System Data Sheet (Figure 6–5).
2. Follow retrofit checklist, steps 1–12 (Figure 6–6).
3. Plot Cycle on MP 39, P/E chart.
4. Measure temperature of evaporator return ells from top to bottom to find glide.

FIGURE 6–5

SYSTEM DATA SHEET

Type of System/Location: _____

    Equipment Mfg.: _____   Compressor Mfg.: _____

    Model No.: _____   Model No.: _____

    Serial No.: _____   Serial No.: _____

    CFC-12 Charge Size: _____   Lubricant Type: _____

    Charge Size: _____

    Drier Mfg.: _____   Drier Type (check one):

    Model No.: _____   Loose fill: _____

    Solid core: _____

Condenser Cooling Medium (air/water): _____

Expansion Device (check one):   Capillary tube: _____

    Expansion valve: _____

    If Expansion Valve:

    Manufacturer _____

    Model No.: _____

    Control/set Point: _____

    Location of Sensor: _____

Other System Controls (ex.: head press control). Describe: _____

_____

(circle units used where applicable)

Date/Time				
Refrigerant				
Charge Size (lb, oz/grams)				
Ambient Temp. (°F/°C)				
Relative Humidity				
Compressor:				
Suction T (°F/°C)				
Suction P (psig, psia/kPa, bar)				
Discharge T (°F/°C)				
Discharge P (psig, psia/kPa, bar)				
Box/Fixture T (°F/°C)				
Evaporator:				
Refrigerant Inlet T (°F/°C)				
Refrigerant Outlet T (°F/°C)				
Coil Air/H_2O In T (°F/°C)				
Coil Air/H_2O Out T (°F/°C)				
Refrigerant T @ Superht. Ctl. Pt. (°F/°C)				
Condenser:				
Refrigerant Inlet T (°F/°C)				
Refrigerant Outlet T (°F/°C)				
Coil Air/H_2O In T (°F/°C)				
Coil Air/H_2O Out T (°F/°C)				
Exp. Device Inlet T (°F/°C)				
Motor Amps				
Run/Cycle Time				

Comments: _____

_____

TABLE 6–1 Pressure-temperature chart. *(Sporlan)*

Notes on the chart: For the zeotropic blends the chart shows both a **BUBBLE POINT** and a **DEW POINT** value. Where two numbers are listed in a single refrigerant column they are given as *bubble / dew*. Values marked with * are in inches mercury below one atmosphere.

PSIG	Pink — MP39 (X) or 401A (X)	Orange — HP62 (S) or 404A (S)	Sand — HP80 (L) or 402A (L)	Teal — AZ-50 (P)	Green — 124 (Q)	Tan — 125
5 *	−23	−57	−59	−58	3	−63
4 *	−22	−56	−58	−57	4	−61
3 *	−20	−54	−56	−56	6	−60
2 *	−19	−53	−55	−54	7	−58
1 *	−17	−52	−54	−53	9	−57
0	−16	−51	−53	−52	10	−56
1	−13	−48	−50	−49	13	−53
2	−11	−46	−48	−47	16	−51
3	−9	−43	−45	−45	18	−49
4	−6	−41	−43	−42	21	−46
5	−4	−39	−41	−40	23	−44
6	−2	−37	−39	−38	26	−42
7	0	−35	−37	−36	28	−40
8	2	−33	−36	−35	30	−39
9	4	−32	−34	−33	32	−37
10	6	−30	−32	−31	34	−35
11	8	−28	−30	−29	36	−33
12	9	−27	−29	−28	38	−32
13	11	−25	−27	−26	40	−30
14	13	−23	−26	−25	41	−29
15	14	−22	−24	−23	43	−27
16	16	−20	−23	−22	45	−26
17	17	−19	−21	−20	46	−24
18	19	−18	−20	−19	48	−23
19	20	−16	−19	−18	49	−22
20	21	−15	−17	−16	51	−20
21	23	−14	−16	−15	52	−19
22	24	−12	−15	−14	54	−18
23	25	−11	−14	−13	55	−16
24	27	−10	−12	−11	57	−15
25	28	−9	−11	−10	58	−14
26	29	−8	−10	−9	59	−13
27	30	−6	−9	−8	61	−12
28	32	−5	−8	−7	62	−11
29	33	−4	−7	−6	63	−10
30	34	−3	−6	−4	65	−8
31	35	−2	−5	−3	66	−7
32	36	−1	−4	−2	67	−6
33	37	0	−2	−1	68	−5
34	38	1	−1	0	69	−4
35	39	2	0	1	71	−3
36	40 / 30	3	0	2	72	−2
37	42 / 31	4	1	3	73	−1
38	43 / 32	5	2	4	74	0
39	44 / 33	6	3	5	75	0
40	45 / 34	7	4	6	76	1
42	46 / 36	8	6	7	78	3
44	48 / 38	10	8	9	80	5
46	50 / 40	12	10	11	82	7
48	42	14	11	13	84	8
50	44	16	13	14	86	10
52	45	17	14	16	88	11
54	47	19	16	17	90	13
56	49	20	18	19	91	15
58	50	22	19	20	93	16
60	52	23	20	22	95	17
62	53	25	22	23	97	19
64	55	26	23	25	98	20
66	56	27	25	26	100	22
68	58	29	26	28	101	23
70	59	30	29 / 27	29	103	24
72	61	32	31 / 29	30	104	26
74	62	33	32 / 30	32	106	27
76	64	34	33 / 31	33	107	28
78	65	35	34 / 32	34	109	29
80	66	37 / 36	34 / 31	35	110	31
85	69	40 / 39	37 / 34	38	114	33
90	73	42 / 42	40 / 37	41	117	36
95	76	45 / 44	42 / 40	44	120	39
100	78	48 / 47	45 / 43	47	123	42
105	81	50	48 / 45	49	126	44
110	84	52	50 / 48	52	129	47
115	87	55	50	54	132	49
120	89	57	53	57	135	51
125	92	59	55	59	138	54
130	94	62	57	61	140	56
135	96	64	60	63	143	58
140	99	66	62	66	145	60
145	101	68	64	68	148	62
150	103	70	66	70	150	64
155	105	72	68	72	152	66
160	108	74	70	74	154	68
165	110	76	72	76	157	70
170	112	78	74	77	159	72
175	114	80	75	79	161	73
180	116	82	77	81	163	75
185	117	83	79	83	165	77
190	119	85	81	85	167	79
195	121	87	82	86	169	80
200	123	88	84	88	171	82
205	125	90	86	90	173	83
210	127	92	87	91	175	85
220	130	95	91	94	178	88
230	133	98	94	97	182	91
240	136	101	97	100	185	94
250	140	104	99	103	188	97
260	143	107	102	106	192	99
275	147	111	106	110	196	103
290	151	115	110	114	201	107
305	155	118	114	118	205	111
320	159	122	118	121	209	114
335	163	126	121	125	213	118
350	167	129	125	128	217	121
365	170	132	128	131	221	124

Temperature, °F — REFRIGERANT (Sporlan Code)

* Inches mercury below one atmosphere

FIGURE 6–6   Checklist for SUVA® MP39 or SUVA® MP66 Retrofit.

_____ **1.** Establish equipment performance with CFC-12. (See data sheet for recommended data.)

_____ **2.** Remove CFC-12 charge from system.
(Need 10–20 in. Hg vacuum (34–67 kPa) to remove charge.)
— Use recovery cylinder/*Do Not Vent to Atmosphere*
— Weigh amount removed (if possible): _____

_____ **3.** Drain lubricant charge from compressor (unless alkylbenzene lubricant is already in compressor).
_____ — Remove 90–95% of lubricant from compressor
— Measure amount removed: _____

_____ **4.** Charge alkylbenzene lubricant.
— Recharge with amount equivalent to amount removed in *Step 3.*

_____ **5.** Reinstall compressor.

_____ **6.** Replace filter drier with new drier approved for use with SUVA MP blends.
—Loose fill driers: use XH9 desiccant
—Solid core driers: check with drier manufacturer for recommendation

_____ **7.** Reconnect system and evacuate with vacuum pump. (Evacuate to full vacuum [30 in. Hg vacuum/0 kPa].)

_____ **8.** Leak check system. (Reevacuate system following leak check.)

_____ **9.** Charge system with SUVA MP39 or SUVA MP66.
—Remove **Liquid Only** from cylinder
—Initial charge 70–75% by weight of original CFC-12 charge
—Amount of refrigerant charged: _____

_____ **10.** Start up equipment and adjust charge until desired operating conditions are achieved.
—Remove **Liquid Only** from cylinder
—If low in charge, add in increments of 3–5% of original CFC-12 charge
—Amount of refrigerant charged: _____
Total Refrigerant Charged (add 9 and 10) _____

_____ **11.** Label components and system for type of refrigerant (SUVA MP39/SUVA MP66) and lubricant (alkylbenzene).

_____ **12.** *Conversion is complete!!*

■ MULTIPLE-CHOICE TEST ■

Name _____

Score _____ Date _____

DIRECTIONS
Circle the letter that best answers the following multiple-choice questions.

1. Flash gas is measured on the:

   a. condensation line
   b. evaporation line
   c. compression line
   d. constant-quality line

2. The following refrigerant is considered nontoxic.

   a. ammonia
   b. sulfur dioxide
   c. chlorodifluoromethane
   d. dichloroethylene

3. Freon refrigerants when exposed to an open flame produce a poison gas called:

   a. phosgene   c. bromine
   b. methylene   d. acetylene

4. The following refrigerant is nonflammable:

   a. propane   c. methylchlorine
   b. butane   d. fluorocarbon

5. A refrigerant cylinder's capacity should never exceed:

   a. 70 percent   c. 90 percent
   b. 80 percent   d. 100 percent

6. Cylinders should never be liquid full at 130°F or:

   a. 44°C   c. 64°C
   b. 54°C   d. 74°C

7. Three identically sized refrigerant cylinders contain variable amounts of liquid R-22 and are exposed to the same ambient temperature. Which one of the following statements is true?

   a. The pressure in the cylinders are equal.
   b. The pressure in the cylinders all vary.
   c. The pressure in one of the cylinders is different from the other two.
   d. The saturated vapor pressures will vary.

8. The highest temperature a refrigerant can exist as a liquid is the:

   a. saturated temperature
   b. critical temperature
   c. absolute temperature
   d. atmospheric temperature

9. "Nonrefillable" on a refrigerant cylinder means:

   a. the cylinder should not be refilled by the user or manufacturers
   b. the cylinder valve will let the refrigerant out but not in
   c. the manufacturer is not permitted to refill the cylinder but the service mechanic may
   d. it is too costly to refill

10. Freon refrigerant leaks can be determined by:

   a. using an electronic leak detector
   b. adding Dytel to the system
   c. using the halide torch
   d. all of the above

■ WRITTEN EVALUATION ■

1. From the information listed in your textbook, classroom discussions, and laboratory activities, explain how to safely handle refrigerants. (Minimum 500 words.)

2. Explain why the pressure-enthalpy diagram is a valuable tool for understanding and analyzing refrigeration performance.

3. If a system has completely lost its refrigerant charge, how would you determine (a) what type of refrigerant the system requires and (b) how the refrigerant escaped?

Unit 7

LOAD CALCULATIONS AND EQUIPMENT SELECTION

■ INTRODUCTION

The tasks in this unit are daily functions of a sales engineer. The sales engineer must first review the plans, measure the job, size the equipment, select the components, and then write a proposal. In order to complete these tasks, you will be required to contact various business establishments for price estimates. Also, you will be required to estimate labor costs.

But because many unforeseen costs can arise before a job is completed or during the warranty period, prices for labor and material may change (and usually in an upward direction). Therefore, in order to ensure a profit of 15 percent, you will probably have to double your cost.

The performance tasks of this unit will require you to make a job survey, draw a floor plan, make an accurate heating and cooling load estimate, and then write a proposal.

Before you attempt a load estimate, make a detailed survey. The estimate should note the load factors and job factors. And because it is very easy to overlook important factors, use standard forms. Several examples have been indicated in the Appendix (see, for example, Weather King Residential Cooling Heating Worksheet, Figure A-2).

The load factors determine the equipment size; the job factors relate to the installation cost. But both factors must be carefully studied in order to spell out what is included in the proposal and who is responsible for the various aspects of the installation.

The next step is to draw a floor plan, such as shown in Figure 7–1. After completing the floor plan, you'll need to know the outdoor design conditions. These temperature design conditions are based on the lowest and highest temperatures recorded for a particular locality. You can consult Figure A-3 for the correct weather data and design conditions. For example, the normal design condition for Los Angeles, California, in July, at 3:00 P.M. is dry bulb, 90°F; wet bulb, 70°F; and average daily range, 14°F.

Once you've completed your job survey sheet and sketched your floor plan, you can complete your heating and cooling estimate sheets (see Figures A-4 and A-5) and be able to select the equipment from the manufacturer's specification sheets.

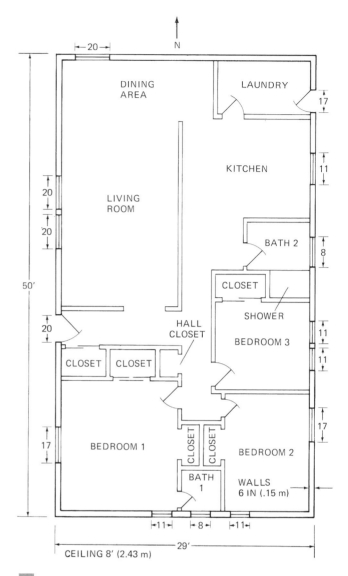

FIGURE 7–1 Floor plan.

■ JOB SHEET 7–1 ■

PERFORMANCE OBJECTIVE
Given a residential floor plan, sufficient survey information, worksheets, and manufacturer's equipment performance data, estimate the heating and cooling load.

REFERENCE
Heating, Ventilating, and Air Conditioning Fundamentals (Chapter 7); sales literature; and Figures 7–1, A-2, A-3, A-4, and A-5

SUPPLIES
Cooling and heating estimate worksheets (Figures A-4 and A-5), notebook

JOB 7-1
Estimating the heating and cooling load

PROCEDURE
1. Complete the job survey sheet of required building data (Figure A-2).
2. Complete the existing installation data sheet.
3. Compile the new installation data.
4. Draw a floor plan (Figure 7–1).
5. Determine the design temperature difference for summer and winter (Figure A-3).
6. Complete the cooling estimate worksheet (Figure A-4).
7. Complete the heating estimate worksheet (Figure A-5).
8. Select the appropriate equipment to match the load estimate sheets.

■ JOB SHEET 7–2 ■

Name _____

Score _____ Date _____

PERFORMANCE OBJECTIVE
Given a residential floor plan, sufficient survey information, worksheets, and manufacturer's equipment performance data, estimate the heating and cooling load.

REFERENCE
Heating, Ventilating, and Air Conditioning Fundamentals (Chapter 7); Figure 7–1; manufacturer's equipment and design temperature data

SUPPLIES
Heating and cooling estimate worksheets (Figures A-4 and A-5)

JOB 7-2
Completing the following heating and cooling worksheet

COOLING ESTIMATE

Step	Procedure	Btu/h	Watt (J/s)
1.	Floor plan (Figure 7–1) 3-in insulation—outside walls	_____	_____
2.	6-in insulation—ceiling	_____	_____
3.	Floors over concrete slab	_____	_____
4.	Unshaded storm windows	_____	_____
5.	Doors	_____	_____
	Three bedrooms	_____	_____
	Appliances	1600	468.6 W
6.	Total Btu/h heat gain for 20° design factor	_____ Btu/h	_____ W
7.	Ductwork with 2-in insulation blanket	_____	_____
8.	NYC, NY, design factor		
	Total Btu/h	_____	_____ W

HEATING ESTIMATE

Step	Procedure	Btu/h	Watt (J/s)
1.	Net exposed walls	_____	_____
2.	6-in insulation—ceiling.	_____	_____
3.	Concrete slab floor, linear ft _____, 0-in insulation	_____	_____
4.	Storm casement windows	_____	_____

5. Weather stripping and
   storm doors   _____ _____
6. Three bedrooms   _____ _____
7. NYC, NY, design factor   _____ _____
8. 2-in duct wrap   _____ _____

■ JOB SHEET 7–3 ■

Name _____

Score _____ Date _____

PERFORMANCE OBJECTIVE
Given a cooling estimate for a specific residence, manufacturer's
specification, and data sheets, select a remote air-cooled condensing unit and a matching A coil evaporator. Equipment must be adequate for Job 7–2 load estimate.

REFERENCE
Heating, Ventilating, and Air Conditioning Fundamentals (Chapter 7); Addison Products Company condensing unit; and evaporator
coil brochures

SUPPLIES
Notebook

JOB 7-3
Selecting an air-cooled condensing unit and a matching A coil

PROCEDURE
1. Cooling estimate (Job 7-2) _____Btu/h
2. Condensing unit data:
   Model: _____
   FLA: _____
   LRA: _____
   EER: _____
   Hard-Start Kit No.. _____
3. A coil evaporator data:
   Model no.: _____
   Outdoor ambient temperature rating: _____ Btu/h
   Airflow cfm: _____

■ TRUE OR FALSE ■

Name _____
Score _____ Date _____

DIRECTIONS

Fill in T for a true statement or F for a false statement.

_____ 1. Blueprints are the main part of the proposal.

_____ 2. Oversized equipment will result in a lower-maintained humidity in the conditioned area.

_____ 3. Oversized equipment can result in short cycling.

_____ 4. Oversized equipment can provide the proper temperature required.

_____ 5. Oversized equipment can result in a higher utility bill for the occupant.

_____ 6. Equipment operating continuously at design conditions is a desirable feature.

_____ 7. A 5 to 7° indoor temperature swing can still provide an average 75°F (23.8°C) indoor temperature.

_____ 8. The indoor temperature swing is usually equal to the outdoor temperature change.

_____ 9. The survey sheet includes load factors and job factors.

_____ 10. The flue size of a furnace that is to be replaced is a job factor.

■ WRITTEN EVALUATION ■

DIRECTIONS

From the worksheet information in Jobs 7-2 and 7-3, secure the prices for similar-type equipment and complete the cost sheet that will include time and material for installation excluding ductwork and electrical work that will be furnished by other contractors. A proposal will be written up following the form provided. Your price quote should be twice the total figure listed on your cost sheet.

Name _____

Score _____ Date _____

Proposal

Contractor: _____

Address: _____

Survey by: _____

Estimate by: _____

Proposal no.: _____

Date: _____

We are pleased to submit our recommendations and quotation for

.... comfort conditioning of your .... located

at .... Our recommendations are in accordance with the engineering practices, standards and certifications of the air-conditioning industry. The equipment and installation specified are recommended to ensure lasting benefit from the most economical and reliable system. A detailed survey and load calculation of the above facility were conducted, and based on these analyses, we are recommending the following major equipment.

MAJOR EQUIPMENT (to be furnished by Seller):

Quantity	Model Number	Description
_____	_____	_____
_____	_____	_____
_____	_____	_____

WARRANTY:

This proposal includes standard manufacturer's warranty applicable to the above equipment at time of purchase.

PRICE:

The total cash price for the equipment and/or installation as described

in this proposal is .... ($.................) including necessary fees, permits, and applicable local taxes.

TERMS:

Payment shall be made as follows: (Complete whichever is applicable)

A—$.........., on acceptance of this proposal by Seller, Balance of the to-

tal cash price in full on (date of)..........(within .... days from date of) completion of installation. B—$.........., cash down payment or trade-in on acceptance of this proposal by Seller, and the Time Balance of

$.......... in .... consecutive monthly installments commencing on the date of completion of installation and on the same day of each

month thereafter. Each installment to be $.........., except the final

installment, which shall be $.......... The finance charge is at annual

percentage rate of .... percent. Buyer to execute and deliver to Seller an installment negotiable note in the amount of the time balance as evidence of the sum owed, and an agreement pursuant to which the title to, the property in, or a lien upon the equipment described in this proposal is retained or taken by the Seller as security, in whole or in part, for the amount owing by the Buyer.

We appreciate very much the opportunity of submitting our quotation, and look forward to your early acceptance so that we may schedule this work for prompt completion, thereby assuring you of the immediate benefit of your investment.

Very truly yours,

By ....

This proposal expires .... days from this date and shall not be binding until accepted in writing by an officer of Seller.

Accepted by (Buyer)   Accepted by (Seller)

. . . .

. . . .   By . . . .

Date . . . .   Date . . . .

Unit 8

PIPING AND LINE SIZING

■ INTRODUCTION

Refrigerant lines, water lines, gas lines, and drain lines must all be of appropriate size to handle the amount of liquid or vapor required. As with calculating heating and cooling loads, calculating the capacities of these lines is important. Faulty piping, undersized lines, and oversized lines cause problems. The system will not operate efficiently. The principles of design are important to every-

one, not only the design engineer. The HVAC mechanic also uses this information to properly diagnose problems caused by faulty piping.

The following tasks will provide practice in using piping tables. A design engineer would be lost without the tables and catalogs that are available from most manufacturers at little or no cost. Everyone in the trade should have piping tables available and know how to interpret them. (See Tables 8–1 to 8–6.)

TABLE 8–1 Recommended suction line sizes, R-22 (40°F [4.4°C] evaporating temperature).

Capacity, Btu/h	Light Load Capacity Reduction, %	Equivalent Length, ft							
		50		100		150		200	
		Horiz.	Vert.	Horiz.	Vert.	Horiz.	Vert.	Horiz.	Vert.
6,000	0	½	½	½	½	⅝	½	⅝	½
12,000	0	⅝	⅝	⅝	⅝	⅞	⅝	⅞	⅝
18,000	0	⅞	⅞	⅞	⅞	⅞	⅞	⅞	⅞
24,000	0	⅞	⅞	⅞	⅞	⅞	⅞	1⅛	⅞
36,000	0	⅞	⅞	1⅛	⅞	1⅛	⅞	1⅛	1⅛
48,000	0	1⅛	1⅛	1⅛	1⅛	1⅛	1⅛	1⅜	1⅛
60,000	0–33	1⅛	1⅛	1⅛	1⅛	1⅜	1⅛	1⅜	1⅛
75,000	0–33	1⅛	1⅛	1⅜	1⅛	1⅜	1⅛	1⅝	1⅜
100,000	0–50	1⅜	1⅜	1⅜	1⅜	1⅜	1⅜	1⅝	1⅜
150,000	0–66	1⅜	1⅜	1⅝	1⅝	1⅝	1⅝	2⅛	1⅝
200,000	0–66	1⅝	1⅝	2⅛	1⅝	2⅛	1⅝	2⅛	1⅝
300,000	0–50	2⅛	2⅛	2⅛	2⅛	2⅛	2⅛	2⅝	2⅛
	66	2⅛	2⅛	2⅛	2⅛	2⅛	2⅛	2⅛	2⅛
400,000	0–66	2⅛	2⅛	2⅛	2⅛	2⅝	2⅛	2⅝	2⅛
500,000	0–66	2⅛	2⅛	2⅝	2⅛	2⅝	2⅛	2⅝	2⅝
600,000	0–66	2⅝	2⅝	2⅝	2⅝	2⅝	2⅝	3⅛	2⅝
750,000	0–66	2⅝	2⅝	3⅛	2⅝	3⅛	2⅝	3⅛	2⅝

TABLE 8–2 Recommended suction line sizes, R-12 (40°F [4.4°C] evaporating temperature).

Capacity, Btu/h	Light Load Capacity Reduction, %	Equivalent Length, ft							
		50		100		150		200	
		Horiz.	Vert.	Horiz.	Vert.	Horiz.	Vert.	Horiz.	Vert.
6,000	0	⅝	⅝	⅝	⅝	⅝	⅝	⅝	⅝
12,000	0	⅞	⅞	⅞	⅞	⅞	⅞	⅞	⅞
18,000	0	⅞	⅞	⅞	⅞	1⅛	⅞	1⅛	1⅛
24,000	0	⅞	⅞	1⅛	1⅛	1⅛	1⅛	1⅛	1⅛
36,000	0	1⅛	1⅛	1⅛	1⅛	1⅜	1⅛	1⅜	1⅜
48,000	0	1⅛	1⅛	1⅜	1⅜	1⅜	1⅜	1⅝	1⅝
60,000	0–33	1⅛	1⅛	1⅜	1⅜	1⅝	1⅜	1⅝	1⅝
75,000	0–33	1⅜	1⅜	1⅝	1⅜	1⅝	1⅜	1⅝	1⅝
100,000	0–50	1⅜	1⅜	1⅝	1⅝	2⅛	1⅝	2⅛	1⅝
150,000	0–33	1⅝	1⅝	2⅛	1⅝	2⅛	1⅝	2⅝	2⅛
	50–66	1⅝	1⅝	2⅛	1⅝	2⅛	1⅝	2⅛	1⅝
200,000	0	2⅛	2⅛	2⅛	2⅛	2⅝	2⅛	2⅝	2⅝
	33–50	2⅛	2⅛	2⅛	2⅛	2⅝	2⅛	2⅝	2⅛
	66	2⅛	2⅛	2⅛	2⅛	2⅛	2⅛	2⅛	2⅛
300,000	0–50	2⅛	2⅛	2⅝	2⅛	2⅝	2⅛	3⅛	2⅝
	66	2⅛	2⅛	2⅝	2⅛	2⅝	2⅛	2⅝	2⅛
400,000	0–50	2⅝	2⅝	3⅛	2⅝	3⅛	2⅝	3⅛	3⅛
	66	2⅝	2⅝	3⅛	2⅝	3⅛	2⅝	3⅛	2⅝
500,000	0–50	2⅝	2⅝	3⅛	2⅝	3⅛	2⅝	3⅝	3⅛
	66	2⅝	2⅝	3⅛	2⅝	3⅛	2⅝	3⅝	2⅝
600,000	0–66	3⅛	2⅝	3⅛	3⅛	3⅝	3⅛	3⅝	3⅛
750,000	0–66	3⅛	3⅛	3⅝	3⅛	3⅝	3⅛	4⅛	3⅝

TABLE 8–3 Allowance for friction loss in valves and fittings.

Fitting Size, in	Equivalent Length of Tube, ft						
	Standard ells		90° tee		Coupling	Gate Valve	Globe Valve
	90°	45°	Side Branch	Straight Run			
⅜	0.5	0.3	0.75	0.15	0.15	0.1	4
½	1	0.6	1.5	0.3	0.3	0.2	7.5
¾	1.25	0.75	2	0.4	0.4	0.25	10
1	1.5	1	2.5	0.45	0.45	0.3	12.5
1¼	2	1.2	3	0.6	0.6	0.4	18
1½	2.5	1.5	3.5	0.8	0.8	0.5	23
2	3.5	2	5	1	1	0.7	28
2½	4	2.5	6	1.3	1.3	0.8	33
3	5	3	7.5	1.5	1.5	1	40
3½	6	3.5	9	1.8	1.8	1.2	50
4	7	4	10.5	2	2	1.4	63
5	9	5	13	2.5	2.5	1.7	70
6	10	6	15	3	3	2	84

TABLE 8–4 Recommended liquid line sizes.

Capacity, Btu/h	R-12					R-22					R-502				
	Condenser to Receiver	Receiver to Evaporator Equivalent Length, ft				Condenser to Receiver	Receiver to Evaporator Equivalent Length, ft				Condenser to Receiver	Receiver to Evaporator Equivalent Length, ft			
		50	100	150	200		50	100	150	200		50	100	150	200
6,000	⅜	⅜	⅜	⅜	⅜	⅜	¼	⅜	⅜	⅜	⅜	¼	⅜	⅜	⅜
12,000	½	⅜	⅜	½	½	½	⅜	⅜	⅜	⅜	½	⅜	½	½	½
18,000	½	½	½	½	½	½	⅜	⅜	½	½	⅝	½	½	½	½
24,000	⅝	½	½	½	⅝	⅝	⅜	½	½	½	⅝	½	⅝	⅝	⅝
36,000	⅝	½	⅝	⅝	⅝	⅝	½	½	½	½	⅞	½	⅝	⅝	⅝
48,000	⅞	½	⅝	⅝	⅞	⅞	½	⅝	⅝	⅝	⅞	⅝	⅝	⅝	⅞
60,000	⅞	⅝	⅝	⅞	⅞	⅞	½	⅝	⅝	⅝	⅞	⅝	⅞	⅞	⅞
75,000	⅞	⅝	⅞	⅞	⅞	⅞	½	⅝	⅝	⅝	⅞	⅝	⅞	⅞	⅞
100,000	1⅛	⅞	⅞	⅞	⅞	⅞	⅝	⅞	⅞	⅞	1⅛	⅞	⅞	⅞	⅞
150,000	1⅛	⅞	⅞	1⅛	1⅛	1⅛	⅞	⅞	⅞	⅞	1⅜	⅞	⅞	1⅛	1⅛
200,000	1⅜	⅞	1⅛	1⅛	1⅛	1⅛	⅞	⅞	1⅛	1⅛	1⅜	1⅛	1⅛	1⅛	1⅛
300,000	1⅝	1⅛	1⅛	1⅜	1⅜	1⅜	1⅛	1⅛	1⅛	1⅛	1⅝	1⅜	1⅜	1⅜	1⅜
400,000	1⅝	1⅜	1⅜	1⅜	1⅜	1⅝	1⅛	1⅛	1⅜	1⅜	1⅝	1⅜	1⅜	1⅜	1⅝
500,000	1⅝	1⅜	1⅜	1⅝	1⅝	1⅝	1⅛	1⅜	1⅜	1⅜	2⅛	1⅜	1⅜	1⅝	1⅝
600,000	2⅛	1⅝	1⅝	1⅝	1⅝	1⅝	1⅜	1⅜	1⅜	1⅝	2⅛	1⅝	1⅝	1⅝	1⅝
750,000	2⅛	1⅝	1⅝	1⅝	2⅛	2⅛	1⅝	1⅝	1⅝	1⅝	2⅛	2⅛	2⅛	2⅛	2⅛

* Maximum delivery capacity in cubic feet of gas per hour of IPS pipe carrying natural gas of 0.65 specific gravity.

Source: Building News, Inc.

TABLE 8–5 Recommended discharge line sizes.

Capacity, Btu/h	Light Load Capacity Reduction, %	R-12 Equivalent Length, ft				R-22 Equivalent Length, ft				R-502 Equivalent Length, ft			
		50	100	150	200	50	100	150	200	50	100	150	200
6,000	0	½	½	½	⅝*	⅜	½	½	½	½	½	½	⅝*
12,000	0	⅝	⅝	⅝	⅞*	½	½	⅝	⅝	⅝	⅝	⅝	⅞*
18,000	0	⅝	⅞	⅞	⅞	⅝	⅝	⅝	⅞	⅝	⅞*	⅞*	⅞*
24,000	0	⅞	⅞	⅞	⅞	⅝	⅞	⅞	⅞	⅞	⅞	⅞	⅞
36,000	0	⅞	⅞	⅞	1⅛	⅞	⅞	⅞	⅞	⅞	⅞	1⅛*	1⅛*
48,000	0	⅞	1⅛	1⅛	1⅛	⅞	⅞	⅞	1⅛*	⅞	1⅛	1⅛	1⅛
60,000	0	1⅛	1⅛	1⅛	1⅜	⅞	1⅛	1⅛	1⅛	1⅛	1⅛	1⅛	1⅜*
	33	1⅛	1⅛	1⅛	1⅜*	⅞	1⅛	1⅛	1⅛	1⅛	1⅛	1⅛	1⅜†
75,000	0	1⅛	1⅛	1⅛	1⅜	⅞	1⅛	1⅛	1⅛	1⅛	1⅛	1⅜	1⅜
	33	1⅛	1⅛	1⅛	1⅜	⅞	1⅛	1⅛	1⅛	1⅛	1⅛	1⅜*	1⅜
100,000	0	1⅛	1⅜	1⅜	1⅝	1⅛	1⅛	1⅜	1⅜	1⅛	1⅜	1⅜	1⅝*
	33–50	1⅛	1⅜	1⅜	1⅝*	1⅛	1⅛	1⅜*	1⅜*	1⅛	1⅜*	1⅜*	1⅝†
150,000	0	1⅜	1⅝	1⅝	2⅛	1⅛	1⅜	1⅜	1⅜	1⅜	1⅜	1⅝	1⅝
	33–50	1⅜	1⅝	1⅝	2⅛*	1⅛	1⅜	1⅜*	1⅜*	1⅜	1⅜	1⅝*	1⅝*
	66	1⅜	1⅝*	1⅝*	2⅛†	1⅛	1⅜*	1⅜*	1⅜*	1⅜*	1⅜*	1⅝†	1⅝†
200,000	0	1⅝	1⅝	2⅛	2⅛	1⅜	1⅜	1⅝	1⅝	1⅜	1⅝	1⅝	2⅛*
	33–50	1⅝	1⅝	2⅛*	2⅛*	1⅜	1⅜	1⅝*	1⅝*	1⅜	1⅝*	1⅝*	2⅛†
	66	1⅝	1⅝	2⅛*	2⅛*	1⅜*	1⅜*	1⅝†	1⅝†	1⅜	1⅝*	1⅝*	2⅛†
300,000	0	2⅛	2⅛	2⅛	2⅛	1⅜	1⅝	1⅝	2⅛	1⅝	2⅛	2⅛	2⅛
	33–50	2⅛	2⅛	2⅛	2⅛	1⅜	1⅝	1⅝	2⅛	1⅝	2⅛	2⅛	2⅛
	66	2⅛*	2⅛*	2⅛*	2⅛*	1⅜	1⅝*	2⅛†	2⅛†	1⅝*	2⅛†	2⅛†	2⅛†
400,000	0	2⅛	2⅛	2⅛	2⅝	1⅝	2⅛	2⅛	2⅛	2⅛	2⅛	2⅛	2⅝
	33–66	2⅛	2⅛	2⅛	2⅝*	1⅝	2⅛*	2⅛*	2⅛*	2⅛*	2⅛*	2⅛*	2⅝†
500,000	0	2⅝	2⅝	2⅝	2⅝	2⅛	2⅛	2⅛	2⅛	2⅛	2⅛	2⅝	2⅝
	33–50	2⅝	2⅝	2⅝	2⅝	2⅛	2⅛	2⅛	2⅛	2⅛	2⅛	2⅝*	2⅝*
	66	2⅝*	2⅝*	2⅝*	2⅝*	2⅛*	2⅛*	2⅛*	2⅛*	Horizontal 2⅝ Double riser 1⅜–2⅛			
600,000	0	2⅝	2⅝	2⅝	3⅛	2⅛	2⅛	2⅛	2⅝	2⅛	2⅝	2⅝	3⅛
	33–50	2⅝	2⅝	2⅝	3⅛*	2⅛	2⅛	2⅛	2⅝*	2⅛	2⅝*	2⅝*	3⅛†
	66	2⅝*	2⅝*	3⅛†	3⅛†	2⅛*	2⅛*	2⅛*	2⅝†	2⅛	2⅝*	2⅝*	3⅛†
750,000	0	3⅛	3⅛	3⅛	3⅛	2⅛	2⅝	2⅝	2⅝	2⅝	2⅝	2⅝	3⅛
	33–50	3⅛	3⅛	3⅛	3⅛	2⅛	2⅝*	2⅝*	2⅝*	2⅝	2⅝	2⅝	3⅛*
	66	3⅛*	3⅛*	3⅛*	3⅛*	2⅛	2⅝*	2⅝*	2⅝*	2⅝*	2⅝*	2⅝*	3⅛†

TABLE 8–6 Size of gas piping*.

Pipe Size, In	Length, ft					
	10	20	30	40	50	60
½	170	118	95	80	71	64
¾	360	245	198	169	150	135
1	670	430	370	318	282	255
1¼	1320	930	740	640	565	510
1½	1990	1370	1100	950	830	760
2	3880	2680	2150	1840	1610	1480

* Maximum delivery capacity in cubic feet of gas per hour of IPS pipe carrying natural gas of 0.65 specific gravity.
Source: Building News, Inc.

■ JOB SHEET 8–1 ■

Name _____

Score _____ Date _____

PERFORMANCE OBJECTIVE
Given a gas piping diagram, and Btu/h requirement for each outlet (gas has specific gravity of 0.65, and 1100 Btu/ft^3), determine the pipe-size requirements for each section and outlet. Sizes should correspond to prerecorded sizes.

REFERENCE
Heating, Ventilating, and Air Conditioning Fundamentals (Chapter 8); Table 8–6, Figure 8–1

SUPPLIES
Notebook, piping diagram (Figure 8–1), tables

JOB 8-1
Finding pipe sizes

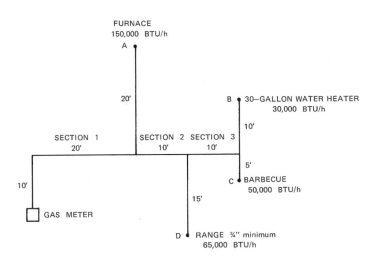

FIGURE 8–1 Gas piping for a typical residence.

PROCEDURE
1. Measure the length of pipe from the gas-meter location to the most remote outlet on the system shown in Figure 8–1.
2. Using Table 8–6, select the column showing that distance. If exact length is not given, choose the next larger distance.
3. Use this vertical column to locate all gas demand figures for this system of piping.
4. Starting at the most remote outlet, find in the vertical column just selected the gas demand for that outlet. If the exact figure of demand is not shown, choose the next larger figure below in the column.
5. The correct size of pipe is the figure opposite the demand figure, in the first column at the left (Table 8–6).

6. Proceed in a similar manner for each outlet and each section of pipe. Determine the gas demand for each section of pipe.

  a. Furnace: _____

  b. Water heater: _____

  c. Barbecue: _____

  d. Gas range: _____

  e. Section 1 demand: _____ft^3/h

  f. Section 2 demand: _____ft^3/h

  g. Section 3 demand: _____ft^3/h

■ JOB SHEET 8–2 ■

Name _____

Score _____ Date _____

PERFORMANCE OBJECTIVE

Given the proper pipe sizing tables and the equivalent length of 200 ft of run, list the AC & R tube sizes for a 150,000 Btu/h condensing unit with a 66 percent light-load-capacity reduction. Sizes should match prerecorded sizes.

REFERENCE

Heating, Ventilating, and Air Conditioning Fundamentals (Chapter 8); Figure 8–2, Tables 8–1, 8–3, 8–4, and 8–5

JOB 8-2

Finding refrigerant line sizing

PROCEDURE

1. The equivalent feet of run is estimated at 200 ft. Select sizes for 40°F (4.45°C) evaporating temperature and 130°F (54.4°C) condensing temperature.
2. Refrigerant R-22 will be used.
3. List the following OD line sizes:
   a. Discharge line: _____
   b. Liquid drain line: _____
   c. Liquid line: _____
   d. Main suction line: _____
   e. Velocity risers: _____

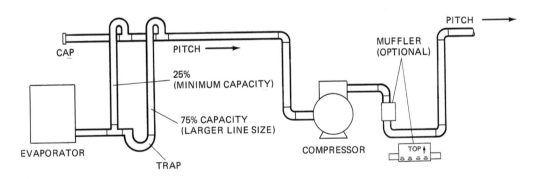

■ **FIGURE 8–2** Double suction riser.

■ JOB SHEET 8–3 ■

Name _____

Score _____ Date _____

PERFORMANCE OBJECTIVE
Given a condenser water piping sketch, desired flow rate and pressure loss through the condenser, and friction-loss charts, select an adequate pipe size and determine the required feet of head for the pump.

REFERENCE
Heating, Ventilating, and Air Conditioning Fundamentals (Chapter 8); Figures 8–3 and 8–4

JOB 8-3
Calculating the required condenser water-line size, using type K copper, 30 gal/min flow rate, and the required feet of head for the pump

PROCEDURE
1. Make a job sketch (see Figure 8–3).
2. Make a pipe and fittings list.
3. Determine the flow rate (30 gal/min).
4. Find the line size for the flow rate (step 3): _____-in ID.
5. Find the equivalent length of pipe and fittings (see Table 8–3).
6. Convert the equivalent length to the pressure drop.
7. Find the condenser pressure drop: _____20 psi for this problem.
8. Find the total pump head: _____ft of head.

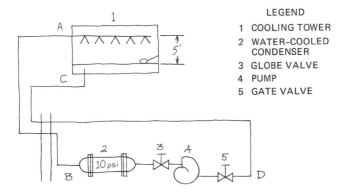

LEGEND
1 COOLING TOWER
2 WATER-COOLED CONDENSER
3 GLOBE VALVE
4 PUMP
5 GATE VALVE

■ **FIGURE 8–3** Sketch of a condenser water circuit.

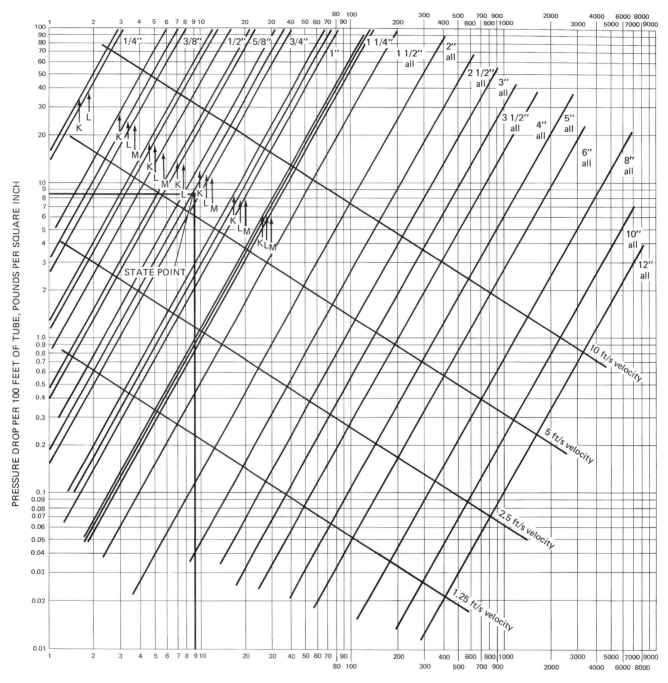

WATER FLOW RATE, GALLONS PER MINUTE
NOTE: Fluid velocities in excess of 5 to 8 ft/s are not usually recommended

FIGURE 8–4   Water flow rate (ID copper tube). *(Copper Development Association, Inc.)*

■ MULTIPLE-CHOICE TEST ■

DIRECTIONS

Circle the letter that best answers the following multiple-choice questions.

1. The refrigerant line that connects the condenser to the receiver is called the:

   a. liquid line   **c.** vapor line
   b. discharge line   **d.** liquid drain line

2. The suction line should be trapped at the evaporator outlet; the riser should:

   a. be sized one size smaller
   b. be sized to lower the velocity requirements
   c. have a slight horizontal pitch
   d. be sized according to the connector size on the unit

3. Unless otherwise specified, AC & R tube-type L is used and the dimensions are given in:

   a. OD (outside diameter)
   b. ID (inside diameter)
   c. newtons/m^2
   d. kPa

4. Undersized refrigerant lines will:

   a. impede the flow of oil back to the compressor
   b. increase the horsepower requirement
   c. lower the horsepower requirement
   d. do *a* and *c* of the above

5. Oversized refrigerant lines result in:

   a. oil problems
   b. low head pressure
   c. lower than normal velocity
   d. all of the above

6. The correct pitch for refrigerant lines is:

   a. $\frac{1}{4}$ in   **c.** $\frac{1}{4}$ in/2.54 cm

   b. $\frac{1}{4}$ in/25.4 mm   **d.** $\frac{1}{4}$ in/30.45 cm

7. The determining factor in sizing a condensate trap is:

   a. the negative static pressure of the fan
   b. the blow-through fan pressure
   c. the drop from the coil to the floor sink
   d. the condensate flow rate

8. Before selecting a water pump and the flow rate (gal/min) you must first determine:

   a. m^3/s   **c.** feet of head
   b. L/s   **d.** horsepower

9. The piping connections on a muffler are not centered with the end caps, therefore:

   a. if it is not positioned in the line properly, it will trap oil
   b. it must always be piped vertically
   c. it should be piped horizontally
   d. the direction of flow is unimportant

10. The normal gas pressure supplied to a residential furnace is 8-in water column or:

    a. 0.5 psi (3.44 kPa)
    b. 0.4 psi (2.75 kPa)
    c. 0.3 psi (2.06 kPa)
    d. 0.2 psi (1.38 kPa)

11. The Btu content of 1 ft^3 of natural gas at 8-in water column and a specific gravity of 0.65 is:

    a. 800 Btu
    b. 900 Btu
    c. 1000 Btu
    d. 1100 Btu

▮ WRITTEN EVALUATION ▮

1. Explain how the job-survey (Unit 7) information can assist the refrigeration fitter.

2. Piping can be run in attics, along walls, near ceilings, and even on the roof. What precautions should be taken to avoid immediate or future problems?

3. Starting with the compressor discharge line of an air-cooled central air-conditioning installation, list the precautionary steps required to ensure oil entrainment and proper lubrication of the compressor.

Unit 9
ELECTRICAL APPLICATION

■ INTRODUCTION

On new construction work the electrician connects the line voltage, runs conduit for the low-voltage controls, and may even mount and connect the thermostat. However, on very few occasions will the electrician also be an HVAC mechanic. Therefore, regardless of who will pull the wire and make the electrical connections, an HVAC mechanic must coordinate the work to make sure everything is properly wired and that the thermostat is in calibration.

The following performance tasks are intended to familiarize you with the low-voltage control circuit and thermostat installation procedures. Also included are performance tasks that evaluate the electrical characteristics of compressor and fan motors.

■ JOB SHEET 9–1 ■

Name _____

Score _____ Date _____

PERFORMANCE OBJECTIVE
Given a low-voltage wall thermostat and sufficient wire to connect the thermostat to the unit, install the thermostat, check for calibration, turn on the power to the unit, and test the thermostat through each stage.

REFERENCE
Heating, Ventilating, and Air Conditioning Fundamentals (Chapter 9); Figures 9–1 through 9–5

EQUIPMENT
Operable air-conditioning unit

TOOLS
Wire stripper, hand tools, small spirit level, plumb line, fish tape, VOM meter

SUPPLIES
No. 16 or 18 AWG cable or stranded wire, wire nuts, notebook

JOB 9–1
Installing a low-voltage thermostat

PROCEDURE
1. Determine the wire size. Use No. 18 AWG cable for runs of less than 50 ft (15.24 m), and No. 16 gauge stranded wire for longer runs.
2. Select the thermostat location. It should be 5 ft (1.5 m) above the floor; in an area with good air circulation; on an inside wall near a return air grill; and away from direct sunlight.
3. Pull about 6 in (15 cm) of low-voltage cable through the wall and plug the hole to prevent drafts. (Drafts may affect the operation.) If the thermostat is out of calibration, follow the manufacturer's recalibration instructions.
4. See Figure 9–1. Pull the low-voltage wiring cable through the square opening shown. Select a plumb line or spirit level and secure subbase to wall (step 3).
5. Match the colors of the wiring cable to the letters on the subbase. Example: See Figures 9–2 and 9–3.

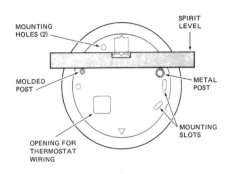

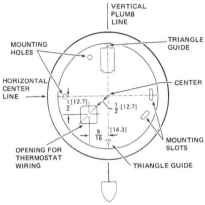

■ **FIGURE 9–1** Installing the subbase.

Color Wire	Subbase	Unit
Orange	O	O–reversing valve
White	W	W–compressor
Green	G	G–fan relay
Red	R	R–transformer

Note: Indicate on paper if substitute colors are used.

6. Turn off the power supply. Warning: Grounding a low-voltage wire or short-circuiting the low-voltage transformer leads can burn out the transformer or blow the control-voltage circuit-protector fuse.

7. Connect thermostat wires to the unit (Figure 9–3). See unit wirings different from heat pumps in Figures 9–4 and 9–5.

8. Turn the thermostat setting 5° below room temperature. Turn selector switch to cooling position.

9. Connect one lead of voltmeter to terminal X (see Figure 9–2) or the common terminal of the low-voltage transformer.

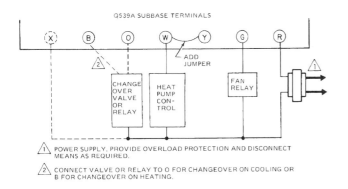

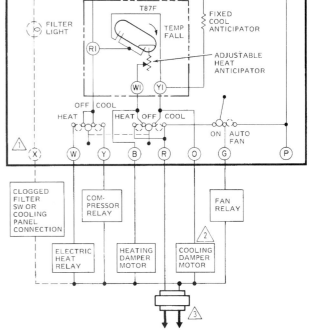

FIGURE 9–2   Heat pump subbase.

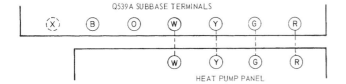

FIGURE 9–3   Wiring heat pump to thermostat.

FIGURE 9–4   Schematic heating-cooling subbase.

10. Twenty-four volts should be read from terminal X to terminals R, Y, O, and G.

11. Set the thermostat 5° above room temperature (call for heating). Place the thermostat selector switch to heating position (Figure 9–4).

12. Twenty-four volts should be read from terminal X to terminal R, W, and B. Note that 24 V will be read to terminal G (fan) only if the fan selector switch (Figure 9–4) is placed in the "on" position.

13. Disconnect power to unit.

14. Remove thermostat cover and observe: Cooling contacts open at room temperature and close at 2° above. Heating contacts close at 2° below room temperature and open at room temperature.

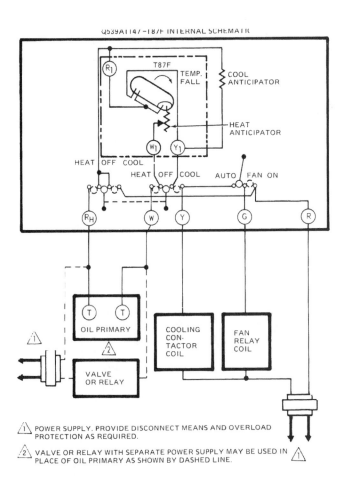

POWER SUPPLY. PROVIDE DISCONNECT MEANS AND OVERLOAD PROTECTION AS REQUIRED.

VALVE OR RELAY WITH SEPARATE POWER SUPPLY MAY BE USED IN PLACE OF OIL PRIMARY AS SHOWN BY DASHED LINE.

FIGURE 9–5   Oil furnace subbase wiring.

■ JOB SHEET 9–2 ■

Name _____

Score _____ Date _____

PERFORMANCE OBJECTIVE
Given an operable air-conditioning unit, solid-state wall thermostat, and 7-day timer, replace an installed electric thermostat with a solid-state thermostat together with a 10-degree night set-back control.

REFERENCE
Heating, Ventilating, and Air Conditioning Fundamentals (Chapter 9); PSG Accustat automatic set-back brochure

EQUIPMENT
AC unit

TOOLS
Hand tools, electrical test equipment

SUPPLIES
Multistage solid-state thermostat with night heat set-back timer, control wiring diagram

JOB 9-2
Installing a solid-state thermostat with automatic set-back control

PROCEDURE
1. Open the unit's electric disconnect switch.
2. Disconnect the existing thermostat.
3. Determine the number of low-voltage wires required.

    ### Example
(1) First stage cool	(5) Transformer
(2) Second stage cool	(6) Night set-back sensor
(3) First stage heat	(7) Fan
(4) Second stage heat	

4. Pull additional wires if needed and connect solid-state thermostat.
5. Mount the time clock on the package unit or en route from thermostat to HVAC components.
6. Select the proper control-wiring diagram (see Figure 9–6).
7. Disconnect the low-voltage wiring from the unit.
8. Connect the wiring in the following manner:

    STAT *to* TIME CLOCK *to* UNIT
STAT	TIME CLOCK	UNIT
N	N	
R	5	R (transformer)
	6	C (transformer)
Y1⟷ Y1	Y1T⟷Y1	
Y2⟷ Y2	Y2T⟷Y2	
W1⟷W1	W1T⟷W1	
W2⟷W2	W2T⟷W2	

9. Set time clock to the correct time.
10. Turn disconnect switch to "on" position and check thermostat.

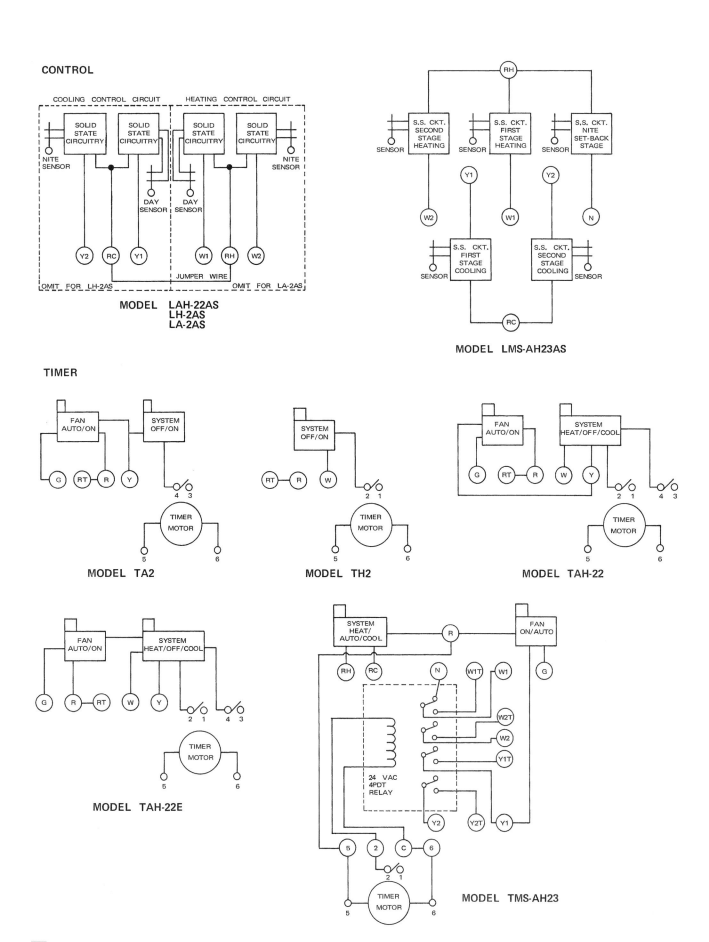

FIGURE 9–6 Wiring of solid-state thermostat.

JOB SHEET 9–3

Name ____________________
Score ____________ Date ________

PERFORMANCE OBJECTIVE
Given a lab air-conditioning unit, remove the condenser fan and complete the necessary paperwork required to obtain the correct replacement motor.

REFERENCE
Heating, Ventilating, and Air-Conditioning Fundamentals (Chapter 9); Figure 9–7, 9–8

EQUIPMENT
Air-cooled remote condensing unit

TOOLS
Hand tools, voltmeter, amp meter

SUPPLIES
Item return tag (Figure 9–7), motor change-out information sheet (Figure 9–8)

JOB 9–3
Completing motor change-out

PROCEDURE
1. Remove the motor from mounts.
2. Remove pulleys, fan blades, or other shaft-mounted items.
3. Complete the "Motor Change-Out Information Sheet" on p. 124.

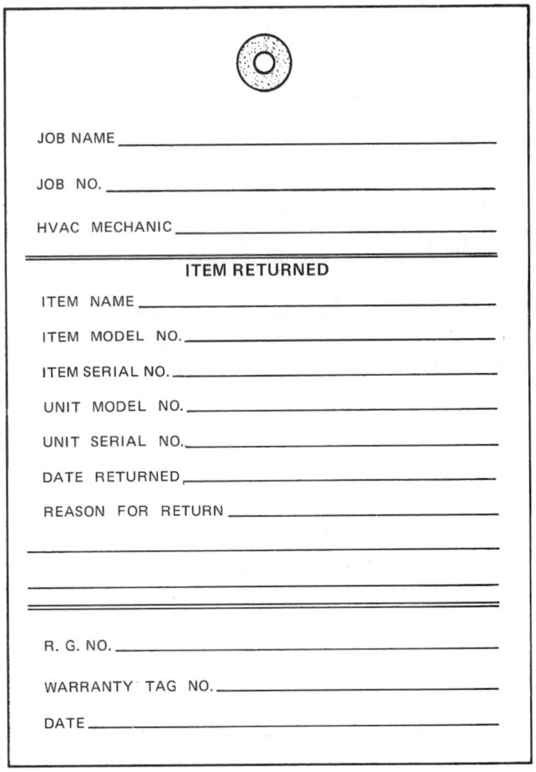
JOB NAME ____________
JOB NO. ____________
HVAC MECHANIC ____________

ITEM RETURNED

ITEM NAME ____________
ITEM MODEL NO. ____________
ITEM SERIAL NO. ____________
UNIT MODEL NO. ____________
UNIT SERIAL NO. ____________
DATE RETURNED ____________
REASON FOR RETURN ____________

R. G. NO. ____________
WARRANTY TAG NO. ____________
DATE ____________

FIGURE 9–7 Item return tag.

4. Call the shop supervisor and give all the information on motor change-out sheet.
5. Complete the item-return-tag information and attach to motor (see p. 123).

HVAC Mechanic: _____   Date: _____

MOTOR CHANGE-OUT INFORMATION

Job Name _____   Job Location _____   Job No. _____

Type Motor: Supply [   ]   Condenser [   ]   Compressor [   ]   Pump [   ]

Motor Mfg. Name _____   Part / Model No: _____

hp _____   rpm _____   Frame No. _____   Voltage and Phase _____

Across the Line   [   ]   or   Increment Start   [   ]   Unit Voltage and Phase _____

Shaft Size _____   Single Shaft [   ]   Double Shaft [   ]   Rotation: CW [   ] or CCW [   ]

Belly Band Mount   [   ]   Resilient Mount   [   ]   Run Capacitor OK ? _____

Open Drip Proof   [   ]   or Totally Enclosed   [   ]

Fan Blade   [   ]   or Blower Wheel   [   ]   OK _____   Width _____ DIA _____

Fan Pitch _____   Number of Blades _____   Rotation: CW [   ] or CCW [   ]

Pulley OK _____   Pulley NO. _____   Pulley OD _____   Bore Size _____

Belt OK _____   Belt Size _____   Contactor/Starter OK _____

Contacts OK _____   Heaters OK _____   Wiring OK _____

Unit Mfr. Name _____   Unit No. (1, 2, 3, etc.) _____

Unit Model No. _____   Unit Serial No. _____

Unit Location _____   Area Served _____

Special Equipment Required _____

_____

_____

FIGURE 9–8   Motor change-out information.

■ JOB SHEET 9–4 ■

Name _____

Score _____ Date _____

PERFORMANCE OBJECTIVE

Given an electronic air cleaner, hand tools, electrical test equipment, neon test lamp for high voltage, and silicon diode:

1. Follow a sequence of checks to locate the cause of failure within an air cleaner.
2. Explain how to locate a faulty component within an assembly or how to prove a component good or bad.

REFERENCE

Heating, Ventilating, and Air Conditioning Fundamentals (Chapter 9); Solid-State Theory, White-Rodgers Division, Emerson Electric; bulletin from Honeywell Electronic Air Filter Service

EQUIPMENT

Screwdrivers—long shank, two plastic or rubber handles; needlenose pliers; test meter with 25-kV dc probe; soldering iron; and neon test lamp for line voltage

SUPPLIES

Silicon diode

JOB 9–4

Troubleshooting a Honeywell electronic air cleaner

PROCEDURE (Diagnostic checks)
1. Energize the electronic air cleaner.
2. Check the collector output.
3. Remove cells.
4. Check the line-voltage power supply and indicator lamp.
5. Isolate the high-voltage power supply from the contact relay.
6. Isolate the transformer from the voltage doubler circuit.

PROCEDURE (Component checks)
1. Make visual inspection.
2. Check for short circuits.
3. Check contact tray and high-voltage wiring.
4. Check for continuity.
5. Check high-voltage transformer.
6. Check voltage-doubler circuit.

■ JOB SHEET 9–5 ■

Name _____

Score _____ Date _____

PERFORMANCE OBJECTIVE
Given an operable air-conditioning unit with a CSR compressor motor, evaluate the performance of the electrical components. Readings must comply with manufacturer's specifications and meet code requirements.

REFERENCE
Heating, Ventilating, and Air Conditioning Fundamentals (Chapter 9).

EQUIPMENT
Operable air-conditioning unit

TOOLS
Hand tools, VOM meter, amprobe, wattmeter

JOB 9–5
Electrical performance evaluation of an AC unit

PROCEDURE

DISCONNECT
1. Check the unit serial plate.
   Is the proper disconnect used? _____
   Wire size: _____AWG
   National Electrical Code requirement: _____AWG
   Fuses: _____Circuit breaker: _____amps

COMPRESSOR
2. Record compressor motor data from serial plate.
   Voltage: _____Frequency: _____
   FLA: _____Service factor: _____
   Horsepower: _____Phase: _____
3. Turn on the unit and record the following:
   Voltage: _____Within code? _____(Y or N)
   Run capacitor(s): _____mfd _____volts
   Start capacitor(s): _____mfd _____volts
   Type start relay: _____Part No. _____
   Overload: _____amps, Part No. _____
   Power factor: _____%
   RLA: _____; LRA: _____
   HP: _____; Brake horsepower: _____

CONDENSER MOTOR

4. Type of motor:
   Split phase _____, Shaded pole _____, Other _____
   Service factor: _____; Power factor: _____%
   Volt rating: _____FLA: _____
   RPM: _____Poles: _____
   Frame: _____Shaft size: _____

SUPPLY FAN

5. Type: _____Power factor: _____
   Volts: _____Phase: _____
   RLA: _____No. speeds: _____

■ MULTIPLE-CHOICE TEST ■

Name _____

Score _____ Date _____

1. In a direct current (dc) circuit the electrons flow:

   a. from a negative to a positive potential
   b. from the positive to the negative terminal
   c. in an alternate direction 60 cycles
   d. parallel to the magnetic lines of force

2. A bimetal thermostat employs a permanent magnet to:

   a. provide snap action
   b. suppress the arc when breaking line voltage contacts
   c. increase the life of the contacts
   d. all of the above

3. Like magnetic fields:

   a. attract
   b. repel
   c. increase the cycles
   d. decrease the cycles

4. Free electrons are more abundant in:

   a. nonconductors
   b. silver
   c. copper
   d. aluminum

5. A three-phase motor requires:

   a. a potential starting relay
   b. a run capacitor
   c. a starting capacitor
   d. no starting components

6. An iron-core transformer with a 240-V primary and 24-V secondary is called a:

   a. high-voltage transformer
   b. step-up transformer
   c. step-down transformer
   d. parallel secondary transformer

7. If the primary coil of a transformer has 208 V and 200 turns, then the secondary coil will have 24 V and:

   a. 25 turns
   b. 50 turns
   c. 75 turns
   d. 100 turns

8. You cannot measure 110 V to ground if the three-phase line is:

   a. delta
   b. wye
   c. star
   d. out of phase

9. A three-pole contactor has:

   a. three contacts
   b. three contacts and overload contacts
   c. three contacts but no overload contacts
   d. three sets of contacts

10. The high-voltage output of an electronic air cleaner is:

    a. less than 400 V ac
    b. over 3000 V ac
    c. less than 400 V dc
    d. over 3000 V dc

■ WRITTEN EVALUATION ■

1. How would you determine if an open low-voltage circuit were due to a faulty thermostat or a broken wire connecting that stat to the unit?

2. Most texts will discuss three types of electromagnet relays: (1) relay, (2) contactor, and (3) magnetic starter. Where are they applied in a pump-down control circuit containing a pilot-duty thermostat with a low-voltage transformer; a 220-V solenoid valve; and a 220-V three-phase compressor motor? The control circuit may contain additional electrical components.

3. If you were to replace a three-phase 5-kW electric heater with a 5-kW single-phase heater of the same voltage, what electrical changes would be required?

Unit 10

SOLID-STATE ELECTRONICS

The following tasks are intended to lead the student through basic troubleshooting procedures associated with solid-state circuitry. Practically every appliance made today has one or more solid-state circuit boards. Upon completion of the following tasks, you will become more familiar with circuit boards, components, and their internal function.

For those who want to go beyond the scope of this chapter or study at home, breadboards and semiconductor devices are readily available through local electronics parts houses. Radio Shack and Heath Company, for example, have inexpensive experimental kits.

The first priority when troubleshooting is a visual inspection. The next step is to verify that ac and dc voltage is present. This involves checking the ac transformer and the dc rectifier. Functional checks of mechanical switches and relays are then made. At this point, if the desired output voltages are not present, the control signal must be traced through the diode, transistor, or integrated circuits.

■ JOB SHEET 10–1 ■

Name _____

Score _____ Date _____

PERFORMANCE OBJECTIVE
Given an electronic circuit with a defective transformer, remove and replace the transformer. The transformer will be securely mounted and will convert input voltage to the proper output voltage.

REFERENCE
Heating, Ventilating, and Air Conditioning Fundamentals (Chapter 10)

TOOLS
Assorted screwdrivers, nut drivers, needle-nose pliers, diagonal pliers, digital VOM, shop towel, soldering iron

SUPPLIES
Resin core solder, soldering braid

JOB 10–1
Replace transformer

PROCEDURE
1. Deenergize system.
2. Mark transformer leads (draw sketch if needed).
3. Disconnect electrical and mechanical connectors.
4. Use solder braid to soak up solder when removing leads from a printed circuit board.
5. Remove defective transformer.
6. Insert replacement transformer.
7. Reconnect mechanical and solder connections.
   Ensure proper polarity when connecting leads.
8. Energize system.
9. Test for performance.
10. Reinstall all access panels.

■ JOB SHEET 10–2 ■

PERFORMANCE OBJECTIVE
Replace a defective power supply on a printed circuit board. The power supply connections will not be loose and they will generate the specified voltage.

REFERENCE
Heating, Ventilating, and Air Conditioning Fundamentals (Chapter 10)

TOOLS
Assorted screwdrivers, needle-nose and diagonal pliers, soldering iron, digital VOM, shop towel

MATERIAL
Resin core solder, soldering braid

JOB 10–2
Replace full wave rectifier power supply

PROCEDURE
1. Deenergize system.
2. Gain access to power supply.
3. Note polarity of connections (make a drawing).
4. Unsolder connections (use braid to remove solder from board).
5. Remove defective power supply.
6. Connect power supply to system (check your drawing for correct polarity).
7. Energize system and check voltages.
8. Replace all access panels that were removed.
9. Check performance.

■ JOB SHEET 10–3 ■

Name _____

Score _____ Date _____

PERFORMANCE OBJECTIVE
Given an electric circuit with a defective switch, record replacement information and replace the switch. The switch must interrupt current when deactivated and restore circuit continuity when activated.

REFERENCE
Heating, Ventilating, and Air Conditioning Fundamentals (Chapter 10)

TOOLS
Assorted screwdrivers, needle-nose and diagonal pliers, soldering iron, digital VOM, shop towel

MATERIAL
Resin core solder, soldering braid

JOB 10–3
Identify, order, and replace various types of switches

PROCEDURE
1. Show the schematic symbol:

2. Voltage rating: _____volts
    Current rating: _____amps.
3. Deenergize system.
4. Gain access to the switch.
5. Make a drawing of connections.
6. Unsolder connections (use braid to remove solder from board).
7. Remove defective switch.
8. Connect switch to system (check your drawing).
9. Energize system and check performance.
10. Deenergize system and replace panels.

■ JOB SHEET 10–4 ■

Name _____

Score _____ Date _____

PERFORMANCE OBJECTIVE
Given an electric circuit with a defective relay, record replacement information and replace the relay. When replaced, the relay must open and close the circuit in accordance with design specifications.

REFERENCE
Heating, Ventilating, and Air Conditioning Fundamentals (Chapter 10)

TOOLS
Assorted screwdrivers, needle-nose and diagonal pliers, soldering iron, digital VOM, shop towel

MATERIAL
Resin core solder, soldering braid

JOB 10–4
Identify, order, and replace a relay

PROCEDURE
1. Deenergize system.
2. Gain access to the defective relay.
3. Draw a schematic symbol of the relay.

4. Voltage rating: _____volts
5. Current rating: _____amps
6. Coil voltage: _____volts
7. Disconnect relay.
8. Remove relay from circuit.
9. Insert replacement relay.
10. Reconnect electrical and mechanical connections.
11. Energize system.
12. Check performance.

■ JOB SHEET 10–5 ■

Name _____

Score _____ Date _____

PERFORMANCE OBJECTIVE
Given an electronic circuit with a defective solid-state diode rectifier, remove rectifier with a low-wattage soldering iron. Observe lead polarity during replacement. Solid-state diode must function to design specifications.

REFERENCE
Heating, Ventilating, and Air Conditioning Fundamentals (Chapter 10)

TOOLS
Assorted screwdrivers, needle-nose and diagonal pliers, soldering iron, heat sinks, digital VOM, shop towel

MATERIAL
Resin core solder, soldering braid

JOB 10–5
Replace solid-state diodes (rectifier)

PROCEDURE
1. Deenergize system.
2. Gain access to the diode.
3. Remove electrical and mechanical connectors.
4. Remove faulty diode (use solder braid).
5. Insert replacement diode.
6. Use heat sinks when soldering leads.
7. Insure lead polarity.
8. Connect mechanical connectors.
9. Energize system.
10. Test for performance.
11. Replace access covers.

■ JOB SHEET 10–6 ■

Name _____

Score _____ Date _____

PERFORMANCE OBJECTIVE

Given an electronic circuit with a defective transistor (SCR, TRIAC), pretest replacement transistor with a transistor tester. There must not be heat or physical damage to the replaced transistor and it must function to specifications.

REFERENCE

Heating, Ventilating, and Air Conditioning Fundamentals (Chapter 10)

TOOLS

Assorted screwdrivers, needle-nose and diagonal pliers, soldering iron, heat sinks, digital VOM, transistor tester, shop towel

MATERIAL

Resin core solder, soldering braid

JOB 10–6

Replace transistor or thyristor (SCR, TRIAC)

PROCEDURE

1. Deenergize the system.
2. Gain access to the transistor.
3. Show the transistor schematic symbol:

4. Disconnect the transistor.
5. Remove the defective transistor.
6. Insert replacement transistor.
7. Do not use excessive heat or form bridges when soldering the leads.
8. Replace access panels.
9. Energize the system.
10. Performance test.

■ MULTIPLE-CHOICE TEST ■

Name _____

Score _____ Date _____

DIRECTIONS
Circle the letter that best answers the following
multiple-choice questions.

1. Assuming the primary windings of a
   transformer has 1,000 turns and 120 volt
   ac, the secondary windings will have 24
   volt output and:

   a. 96 turns   **c.** 240 turns
   b. 200 turns   **d.** 2000 turns

2. The best instrument for testing a transformer is:

   a. VOM   **c.** wattmeter
   b. amprobe   **d.** ohm meter

3. Fuse protection for a 40 VA, 24-volt transformer is:

   a. 1.33 amp   **c.** 1.66 amp
   b. 1.5 amp   **d.** 2 amp

4. How many diodes does a bridge rectifier
   employ?

   a. 1   **b.** 2   **c.** 3   **d.** 4

5. What is the minimum amount of diodes
   for a full-wave rectifier?

   a. 1.   **b.** 2   **c.** 3   **d.** 4

6. Looking at a schematic symbol for a
   diode, current flows:

   a. toward the arrow   **c.** either way
   b. with the arrow   **d.** not applicable

7. What test instrument do you use to test a
   switch?

   a. LED   **c.** wiggy
   b. VOM   **d.** amprobe

8. An on/off switch is

   a. SPST   **c.** two pole
   b. SPDT   **d.** 1PNC

9. How many connections does a SPDT
   switch have?

   a. 1   **b.** 2   **c.** 3   **d.** 4

10. How many contacts does a single-pole
    relay have?

    a. 1   **b.** 2   **c.** 3   **d.** 4

11. A relay is what type of switching device?

    a. electromechanical   **c.** magneto
    b. solenoid   **d.** on/off

12. What type of solder is used for electron-
    ics?

    a. acid core   **c.** 95-5
    b. resin core   **d.** 50-50

13. Which element is not an element of a
    transistor?

    a. anode   **c.** collector
    b. base   **d.** emitter

14. The purpose of a soldering braid is to:

    a. absorb heat
    b. absorb solder
    c. shield heat
    d. physically protect

15. The heat sink:

    a. prevents cold solder joints
    b. assists heat transfer
    c. absorbs heat away from transistor
    d. solidifies the solder joint

▨ WRITTEN EVALUATION ▨

1. What precautions are necessary when working on PC boards?
2. Relate the purpose of a power supply.
3. When do you need a heat sink?
4. What should you look for when making a visual inspection of a PC board?

Unit 11

GAS-FIRED FURNACES

■ INTRODUCTION

The following tasks are directed toward fuel conservation. Fuel conservation begins with pilot-burner control. The pilot should be lit only when there is a call for heating, which requires an electric, or electronic, ignition system in addition to the proper air and pressure adjustments.

The second step to fuel conservation is to ensure complete fuel combustion while burning only the minimum amount of fuel necessary to provide comfort. This can be accomplished by installing an electronic thermostat with an electric time clock, as described in Chapter 9 of the text. The solid-state stat is highly sensitive to tempera-

ture change; therefore, it can maintain a fairly constant room temperature. The time clock lowers the night set point and can cut fuel requirements up to 50 percent (see Figure 11–1).

Locking covers also are available for all types of thermostats. Figure 11–2 shows the round Honeywell thermostat with a locking cover which is manufactured by Watsco Products. Locking covers prevent unauthorized persons from changing the temperature settings.

The following performance tasks are designed to help you develop skills of pilot adjustment, burner adjustment, and thermostat adjustment that includes night setback and heat-anticipator setting.

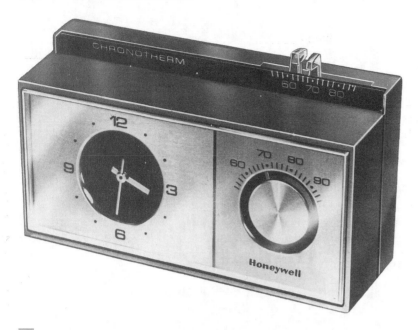

FIGURE 11–1 Night setback thermostat.

FIGURE 11–2 Locking thermostat cover.
(Watsco Products)

■ JOB SHEET 11–1 ■

PERFORMANCE OBJECTIVE
Given a gas-fired furnace with a pilot safety device, light the pilot and adjust the flame to specifications listed in the text.

REFERENCE
Heating, Ventilating, and Air Conditioning Fundamentals (Chapter 11)

TOOLS
Hand tools, two pipe wrenches

SUPPLIES
Matches (match holder if needed)

JOB 11–1
Lighting the pilot and adjusting the flame

PROCEDURE
1. Turn the gas line shut-off valve at unit to the "off" position.
2. Disconnect the union downstream of shut-off valve.
3. Crack open the shut-off valve to purge the air from the gas-supply line.
4. Tighten the union and reopen the gas line.
5. Remove the access panels from the furnace.
6. Turn the gas valve to pilot position and compress pilot drop-out spring to the operating position.
7. Light the pilot.
8. If the flame is yellow, check the primary air hole on the pilot burner and clean if necessary. (Sometimes the air hole is clogged or restricted with a spider web.)
9. Turn the pilot flame adjustment screw so that the flame only contacts the last ½ in (12.5 mm) of the thermal couple or pilot generator. Turn the gas valve to "on" position.
10. Use a soap-bubble solution to check gas connections for leaks.
11. Reinstall the pilot access panel.

■ JOB SHEET 11–2 ■

Name _____

Score _____ Date _____

PERFORMANCE OBJECTIVE
Given a gas burner and screwdriver, adjust the burner to produce a
soft blue flame with the base of the flame not leaving the burner.

REFERENCE
Heating, Ventilating, and Air Conditioning Fundamentals (Chapter 11)

EQUIPMENT
Gas-fired furnace

TOOLS
Hand tools

SUPPLIES
Jumper wire with alligator clips, matches (match holder if needed)

JOB 11–2
Adjusting the gas burner to produce a proper flame

PROCEDURE
1. Remove the burner access panels.
2. Set the thermostat to call for heating or connect a jumper
   wire from transformer terminal R to terminal W1.
3. Observe the flame:
   (a) Hard blue flame—too much primary air.
   (b) Yellow flame—insufficient primary air.
   (c) Soft blue flame—correct air to gas mixture.
4. Adjust the primary air shutter to produce a soft blue flame
   that does not leave the burner.
5. Do not omit this test! With the main burner in operation,
   paint pipe joints, pilot gas tubing connections, and valve
   gasket lines with a rich soap-and-water solution. Bubbles indicate gas leakage. To stop a leak, tighten the joints and
   screws or replace gasket.
6. Reset thermostat (remove jumper).
7. Replace access panels.

■ JOB SHEET 11–3 ■

Name _____

Score _____ Date _____

PERFORMANCE OBJECTIVE
Given a chronotherm thermostat, set the temperature control, set the timer to lower the heating set point for a specified period of time every 24 hours, and set the heat anticipator.

REFERENCE
Heating, Ventilating, and Air Conditioning Fundamentals (Chapter 11); "Chronotherm Service Bulletin," Honeywell (Figure 11–1)

EQUIPMENT
Chronotherm thermostat

TOOLS
Hand tools

JOB 11–3
Setting a chronotherm thermostat

PROCEDURE
1. Remove cover and use clock set-wheel to move clock hands to the correct time. Caution: Do not move the clock hands or turn the timer dial to set the clock.
2. The timer dial has a day and night portion. The time indicator must point to the correct half of the timer scale.
3. The timer has been adjusted at the factory to lower the temperature setting at 10:30 P.M. and raise the temperature setting at 6:00 A.M. The minimum period for high setting is 5½ hours; the minimum period for low setting is 4½ hours. To change the setting, remove the thermostat cover.
4. To change the "Lo" setting, push the tab marked "Lo" inward. Then move it to the time you want the temperature to be lowered. Do not attempt to turn the dial.
5. To change the "Hi" setting, push the tab marked "Hi" inward. Then rotate to the desired time.
6. Check the current draw of the primary control or gas valve(s). The current rating is usually stamped on the nameplate of the device. There also may be a heat-anticipator setting listed in the manufacturer's instructions. Move the pointer on the scale to correspond to this rating.
7. If two gas valves are used (high flame/low flame), the two-stage stat will have an anticipator for each solenoid and the current draw is usually different. Check and set the anticipators accordingly.
8. Replace the thermostat cover.
9. The nonlocking cover can be replaced with a locking cover without disconnecting the thermostat.

■ JOB SHEET 11–4 ■

Name _____

Score _____ Date _____

PERFORMANCE OBJECTIVE
Given the proper tools and test instruments, start up a gas-fired furnace. Heat anticipators will match current draw of gas valve, and burners will be adjusted for complete combustion.

REFERENCE
Heating, Ventilating, and Air Conditioning Fundamentals (Chapter 11)

TOOLS
Hand tools, two pipe wrenches, amprobe, flash light, manometer

MATERIAL
Pipe dope

JOB 11–4
Start up gas fired furnace

PROCEDURE
1. Open and adjust supply and return air registers.
2. Set thermostat for ventilation.
3. Jumper fan switch
4. Supply fan FLA: _____amps; RLA: _____amps
5. Read current draw rating stamped on gas valve: _____amps
6. Set heat anticipators on thermostat (step 5 amps).
7. Purge gas supply line of air.
8. Check all wiring connections.
9. Is cord properly polarized? _____
10. Are cabinet ground wires (green wire) connected?
11. Remove jumper wire and set fan switch: _____°F.
12. Check high limit setting: _____°F.
13. Set thermostat to call for heat.
14. Check ignition.
15. Check burner combustion air adjustment.
16. Describe flame: _____.
17. Measure burner gas pressure: _____in Wc.

■ MULTIPLE-CHOICE TEST ■

Name _____

Score _____ Date _____

DIRECTIONS
Circle the letter that best answers the following multiple choice questions.

1. The specific gravity of natural gas is:

   a. 0.6   **c.** 1.6
   b. 1   **d.** 2

2. The most commonly used fuel in the United States is:

   a. wood
   b. coal
   c. liquefied petroleum gas
   d. natural gas

3. When a gas-fired unit is installed in a closet:

   a. a vent is needed for combustion air make-up
   b. the closet door must have a louver with a free area size of 1 in^2 per 1000 Btu/h
   c. the room must have a vent near the ceiling height and a lower vent near the floor level
   d. additional ventilation is unnecessary if the gas-fired unit has a powered vent motor

4. Primary burner adjustment is properly set at:

   a. 5 parts air to 1 part gas
   b. 10 parts air to 1 part gas
   c. 5 parts gas to 1 part air
   d. 10 parts gas to 1 part air

5. If the heat exchanger is restricted with carbon deposits (soot):

   a. the flame will be blue
   b. the flame will be yellow

   c. the flame is receiving more primary air than secondary air
   d. the secondary air is overdiluted

6. The vent pipe must have a minimal horizontal rise of ½ in per ft or:

   a. 10 mm per meter
   b. 20 mm per meter
   c. 40 mm per meter
   d. 80 mm per meter

7. Natural gas for residential use is reduced in pressure at the meter to:

   a. 10 psi (68.9 kPa)
   b. 55 psi (379 kPa)
   c. 8.5 in (2.1 kPa) water column
   d. 3.5 in (871 pa) water column

8. A safety pilot controlled by a single thermocouple will drop out at:

   a. 4 mV
   b. 8 mV
   c. 16 mV
   d. 25 to 35 mV

9. The thermocouple that can automatically operate a diaphragm-type gas valve is commonly called:

   a. pilot generator   **c.** pilot orifice unit
   b. pilot stat   **d.** solenoid coil

10. The heat-anticipator setting on a heating stat must match the solenoid gas valve's:

    a. coil resistance
    b. watt rating
    c. ampere rating
    d. voltage rating

■ WRITTEN EVALUATION ■

1. The existing job has a constant pilot. What is required to change the unit to either electric or electronic ignition?
2. The pilot gas is on, but the main burner will not light. How do you go about solving the problem?
3. A wall heater has a manual shut-off valve, but no safety pilot or thermostat. List the parts and procedure for converting the heater to an automatic millivolt system.

Unit 12

OIL-FIRED FURNACES

INTRODUCTION

After installation of an oil furnace is completed, the next step to follow is the start-up or final checkout. There are several problems you may be confronted with at this point. The burner, for example, may not start, or the burner may start but not establish a flame. The accompanying Blueray Troubleshooting Guide (Table 12–1) can help you find the solution to your problem.

For optimum performance, follow the manufacturer's specifications. (See Table 12–2.) These specifications can be used as a checklist. For example, we see in Table 12–2 that the Blueray model BF-60 furnace requires a 20 in × 25 in × 1 in filter. This serves as a reminder to make sure the proper filter is installed.

The first two performance objectives are steps you should complete after every installation. The first objective covers start-up procedure. The second objective is a service follow-up procedure. All installations should be reinspected after one or two weeks of normal operation. The unit could have been installed in the summer, at which time a good checkout may not have been practical. The second step can also serve as a preventive maintenance guide or be classified as a preseason tune-up.

The final task involves servicing the burner. It can be used as a guide. However, because electrode placement and settings vary, check the installation and service instructions that were furnished by the manufacturer for a particular burner. The following burner servicing instructions are recommended for Blueray burners

TABLE 12–1   Blueray troubleshooting guide.

Trouble: Burner Does Not Start

Source	Procedure	Causes	Remedy
Thermostat	Check thermostat settings.	Thermostat set too low.	Turn thermostat up.
		Thermostat on off or cool.	Switch to heat.
	Jump TT terminals on primary control. If burner starts, fault is thermostat circuit.	Open thermostat wires.	Repair or replace wires.
		Loose thermostat connectors.	Tighten connection.
		Faulty thermostat.	Replace thermostat.
		Thermostat not level.	Level thermostat.
		Dirty thermostat contacts.	Clean contacts.
Circuit overloads	Check burner motor overload switch.	Burner motor tripped on overload.	Push reset button.
	Check primary control safety switch.	Primary tripped on safety.	Reset safety switch.
Power	Check furnace disconnect switch and main disconnect switch.	Switch open.	Close switch.
		Tripped breaker or blown fuse.	Reset breaker or replace fuse.
Pyrostat	Jump the FF terminals on primary control. If the burner starts, fault is in detector circuit.	Open pyrostat wires.	Repair or replace wires.
		Detector contacts out of step.	Place detector contacts in step.
		Faulty pyrostat.	Replace pyrostat.
Primary control	Check for line voltage between the black and white leads. No voltage indicates no power to the control.	Blower control switch open.	Check limit setting (200°F).
			Jump terminals: if burner starts replace control.
		Open circuit between blower control and disconnect switch.	Repair circuit.
		Low line voltage or power failure.	Call utility company.
	Check for line voltage between orange and white leads. No voltage indicates a fault control.	Defective control.	Replace control.
Burner	Check for voltage at the black and white leads to the burner motor. Voltage indicates power to motor and a fault in the burner.	Fuel pump seized.	Turn off power to burner.
		Blower wheel binding.	Rotate blower by hand to check for excessive drag. Replace fuel unit or blower wheel.
		Burner motor defective.	Replace burner motor.

Trouble: Burner Starts but Does Not Establish Flame

Oil supply	Check tank for oil.	Empty tank.	Fill tank.
	Check for water in oil tank using a dipstick coated with litmus paste.	Water in oil tank.	Strip tank of water exceeding 2 in in depth.
	Listen for pump whine.	Fuel supply valve closed.	Open valve.
Oil line and filter	Open pump bleed port and start burner. Milky oil or no oil indicates loss of prime.	Air leak in fuel system.	Repair leak. Use only flared fittings. Do not use teflon tape on oil fittings.

■ **TABLE 12–1**   *Continued*

Source	Procedure	Causes	Remedy
	Listen for pump whine.	Oil filter plugged.	Replace filter cartridge.
		Plugged pump strainer.	Clean strainer.
		Restriction in oil line.	Repair oil line.
Oil pump	Install pressure gauge in port of fuel pump. Pressure should be 100 psi.	Pump worn—low pressure. Motor overloads.	Replace pump.
		Coupling worn or broken.	Replace coupling.
		Pump discharge pressure set too low.	Set pressure at 100 psi.
Air metering plate	Check for loose play by applying pressure to buss bars.	Air-metering plate not driven up tightly to end of blast tube.	Loosen thumbnut. Drive metering plate assembly up tight. Secure thumbnut.
	Check for correct metering plate specifications. Plate specifications should match firing rate of furnace.	Incorrect metering plate installed.	Install metering plate stamped for firing rate specified for furnace.
Ignition electrodes	Remove air metering plate. Assemble and inspect electrodes and buss bars.	Carboned and shorted electrodes.	Clean electrodes.
		Eroded electrode tips.	Dress up tips and reset electrodes.
		Incorrect electrode settings.	
Ignition transformer	Connect transformer leads to line voltage. Listen for spark. Check that transformer terminals are not arcing with buss bars. Check that transformer is properly grounded.	No spark or weak spark.	Replace transformer.
		Line voltage below 102 V.	Call utility company.
Burner motor	Burner motor trips on overload. Turn off power and rotate blower by hand to check for excessive drag.	Line voltage below 102 V.	Call utility company.
		Faulty motor.	Replace motor.
		Pump or blower overloading motor.	Replace a pump or blower.
Nozzle	Inspect nozzle for plugged orifice and distributor slots.	Plugged orifice or distributor.	Replace nozzle with nozzle specified on burner housing.
		Plugged nozzle strainer.	
		Poor spray pattern.	
	Inspect nozzle for correct size and specifications.	Incorrect nozzle installed.	

Trouble: Burner Fires, but then Fails on Safety

Pyrostat	After burner fires open pyrostat circuit if flame looks OK. If burner continues to operate, fault is in pyrostat.	Faulty pyrostat.	Replace pyrostat.
Primary control	After burner fires, open pyrostat circuit if flame looks OK. If burner locks out, fault is in primary control.	Faulty primary control.	Replace primary control.

■ **TABLE 12–1** *Continued*

Source	Procedure	Causes	Remedy
Poor fire	Inspect flame for shape and uniformity of color.	Unbalanced fire.	Replace nozzle with specified nozzle.
		Excessive draft.	Reduce draft setting.
		Insufficient draft.	Increase draft.
		Air metering plate not driven up tightly to end of blast tube.	Loosen thumbnut. Drive metering plate assembly up tight. Secure thumbnut.
		Incorrect air metering plate installed.	Install metering plate stamped for firing rate specified for furnace.
		Too little combustion air.	Increase combustion air.
Heat-exchanger restriction	Take draft reading at flue box and read draft over the fire with a long probe inserted through the heat exchanger tube. Difference should not exceed 0.01 in.	Plugged heat exchanger.	Clean out heat exchanger.

Trouble: Burner Fires, but then Loses Flame

Source	Procedure	Causes	Remedy
Poor fire	Inspect flame for stability.	Unbalanced fire.	Replace nozzle with specified nozzle.
		Excessive draft.	Reduce draft setting.
		Insufficient draft.	Increase draft.
		Air metering plate not driven up tightly to end of blast tube.	Loosen thumbnut. Drive metering plate assembly up tight. Secure thumbnut.
		Incorrect air metering plate installed.	Install metering plate stamped for firing rate specified for furnace.
		Too little combustion air.	Increase combustion air.
Oil supply	If burner loses flame prior to the primary control locking out, fault is in fuel system.	Air leak in fuel system.	Repair leak; use only flared fittings.
		Water in oil tank.	Strip tank of water exceeding 2 in in depth.
		Fuel-supply valve closed.	Open valve.
		Restriction of oil line.	Clean oil-line restriction.
		Plugged fuel filter.	Replace filter cartridge.
		Plugged pump strainer.	Clean strainer.
		Cold oil.	Use #1 heating oil.

■ **TABLE 12–1** *Continued*

Source	Procedure	Causes	Remedy

Trouble: Burner Fires, but Operates with Low CO_2

Source	Procedure	Causes	Remedy
Combustion air	Reduce combustion-air supply.	Too much combustion air.	Close air band and air shutter to raise CO_2.
Air metering plate	Check for loose play by applying pressure to buss bars.	Air metering plate not driven up tightly to end of blast tube.	Loosen thumbnut. Drive metering plate assembly up tight. Secure thumbnut.
	Check for correct metering plate specifications. Plate specifications should match firing rate of furnace.	Incorrect metering plate installed.	Install metering plate stamped for firing rate specified for furnace.
Pump	Install pressure gauge in gauge port of fuel pump. Pressure should be 100 psi.	Pump discharge pressure incorrectly set.	Set pressure at 100 psi.
		Coupling worn or broken.	Replace coupling.
		Pump worn—low pressure motor overloads.	Replace pump.
Excessive draft	Take a draft reading. Draft should be 0.01 to 0.02 in water column.	Incorrect draft setting.	Reduce draft setting. Install second draft regulator if necessary.
Poor flue gas sample	Insert CO_2 probe into heat exchanger tube. If reading is greater by ½ percent or more, sample was being diluted near flue box.	Leak in flue system.	Sample CO_2 in heat exchanger. Seal flue-system leak.
Testing method	Using a chemical absorption-type device, let instrument set after a test before venting. If CO_2 reading increases ½ percent, fluid is weak.	Weak fluid.	Replace fluid in testing device.
Nozzle	Inspect nozzle for plugged orifice and distributor slots.	Plugged orifice or distributor.	Replace nozzle (cc only) with nozzle specified on burner housing.
		Plugged nozzle strainer.	
		Poor spray pattern.	
Heat exchanger leaking	Take CO_2 readings after unit is warm with the blower on and blower off. Take blower off reading between time burner starts and blower starts. A difference exceeding 1 percent usually indicates a heat-exchanger leak.	Leaking gasket.	Replace gasket.
		Corroded heat exchanger.	Repair or replace heat exchanger.

Trouble: Burner Fires but Pulsates

Source	Procedure	Causes	Remedy
Draft	Take a draft reading. Draft should be 0.01 to 0.02 in.	Down drafts.	Install vent cap.
		Insufficient draft.	Increase draft setting.
		Excessive draft.	Reduce draft setting. Install second draft regulator if necessary.

■ **TABLE 12–1**   *Continued*

Source	Procedure	Causes	Remedy
Draft regulator	Inspect draft regulator for correct location on flue system.	Improper installation.	Move draft regulator to correct location.
Combustion air	Inspect installation for combustion-air provisions.	Improper installation.	Provide openings that freely communicate with outside.
	Open air band wide and take CO_2 reading.	Improper adjustment.	Adjust CO_2 level; start with the air band wide open.
Temperature rise	Measure the temperature rise across the heat exchanger. Rise should not exceed 90°F.	Insufficient air movement over heat exchanger.	Increase blower speed; increase duct sizes.
			Replace fouled air filter.
Oil supply	Bleed pump; inspect for air leaks or water contamination.	Air leak in fuel system.	Repair leak; use only flared joints.
		Water in oil tank.	Strip tank of water exceeding 2 in in depth.
Pump pressure	Install pressure gauge in gauge port of fuel pump. Pressure should be 100 psi.	Pump discharge pressure incorrectly set.	Set pressure at 100 psi.
		Coupling worn or broken.	Replace coupling.
		Pump worn; low-pressure motor overloads.	Replace pump.
Nozzle	Inspect nozzle for plugged orifice and distributor slots.	Plugged orifice or distributor.	Replace nozzle with nozzle specified on burner housing.
		Plugged nozzle strainer.	
		Poor spray pattern.	
Air metering plate	Check for loose play by applying pressure to buss bars.	Air metering plate not driven up tightly to end of blast tube.	Loosen thumbnut. Drive metering plate assembly up tight. Secure thumbnut.
	Check for correct metering plate specifications. Plate specifications should match firing rate of furnace.	Incorrect metering plate installed.	Install metering plate stamped for firing rate specified for furnace.
Heat-exchanger restriction	Take draft reading at flue box and read draft over the fire with a long probe inserted through the heat exchanger tube. Difference should not exceed 0.01 in.	Plugged heat exchanger.	Clean out heat exchanger.
Heat-exchanger leaking	Take CO_2 readings after unit is warm with the blower on and blower off. Take blower off reading between time burner starts and blower starts. A difference exceeding 1 percent usually indicates a heat-exchange leak.	Leaking gasket.	Replace gasket.
		Corroded heat exchanger.	Repair or replace heat exchanger.

Source: Blueray Systems, Inc.

▪ TABLE 12–2 Specifications.

Model	Heating		Cooling	
	BF-60	*BF-75*	*BF-60*	*BF-75*
Output,* (Btu/h)	70,000	88,000	70,000	88,000
Input, Btu/h	84,000	105,000	84,000	105,000
Nozzle (Monarch cc)	0.60–70°	0.75–70°	0.60–70°	0.75–70°
Air metering plate	0.60B	0.75B	0.60B	0.75B
Stack draft (water column)	0.01 in N	0.01 in N	0.01 in N	0.01 in N
CO_2 (nominal)	13.0%	13.0%	13.0%	13.0%
Smoke	0	0	0	0
Net stack temperature, °F	470°	450°	470°	450°
Heat transfer surface, in²	2374	2948	2374	2948
Combustion chamber diameter, in	9⅛	10⅝	9⅛	10⅝
Heat exchanger material (gauge)	16	16	16	16
Heat exchanger tubes (number)	12	16	12	16
Flue diameter in	5	5	5	5
Fuel pump†	A2VA7016	A2VA7016	A2VA7016	A2VA7016
Pump pressure, psi	100	100	100	100
Burner motor‡ hp—rpm	1/7–3450	1/7–3450	1/7–3450	1/7–3450
Flame sensor (Honeywell)	C550D1005	C550D1005	C550D1005	C550D1005
Primary relay (Honeywell)	R8189D1007	R8189D1007	R8189D1007	R8189D1007
Thermostat heat anticipator, A	0.2	0.2		
Blower CFM	730	910	1000	1050/1270
Blower RPM	900	860	1100	1010/1120
Blower motor tap	Low No. 4	Low No. 4	High No. 2	H. No. 2/M. No.3
Static pressure (water column)	0.15	0.20	0.42	0.50
Blow wheel, in	9 × 7	10 × 7	9 × 7	10 × 7
Blower motor (3-speed D.D.)	⅕ hp	⅓ hp	¼ hp	⅓ hp
Maximum air-conditioning capacity, tons	—	—	2½	3
Cabinet width, in	15½	15½	15½	15½
Cabinet depth, in	34½	34½	34½	34½
Cabinet height, in	45½	45½	45½	45½
Flue height C/L, in	39½	39½	39½	39½
Supply plenum width, in	14½	14½	14½	14½
Supply plenum depth, in	21½	21½	21½	21½
Return plenum width, in	19	19	19	19
Return plenum depth, in	24	24	24	24
Filter size, in	20 × 25 × 1	20 × 25 × 1	20 × 25 × 1	20 × 25 × 1
Approximate shipping weight, lb	200	215	200	215

*Based on 83 percent combustion efficiency.
†Sundstrand fuel unit or equivalent.
‡1/8-hp motor suitable for replacement.
Source: Blueray Systems, Inc.

■ JOB SHEET 12–1 ■

PERFORMANCE OBJECTIVE
Given an oil-fired furnace, manufacturer's unit specifications, and portable combustion test kit, perform a final checkout.

REFERENCE
Heating, Ventilating, and Air Conditioning Fundamentals (Chapter 12); Blueray Troubleshooting Guide (Table 12–1); unit specifications, combustion test kit instructions (Dwyer Instruments)

EQUIPMENT
Oil-fired furnace

TOOLS
Hand tools, Dwyer portable combustion test kit

SUPPLIES
Smoke test paper

JOB 11–1
Performing a final checkout

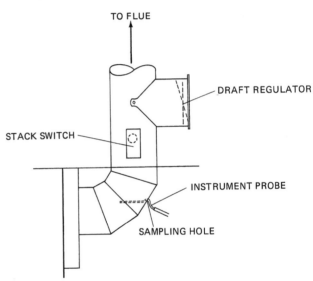

■ FIGURE 12–1 Draft regulator and sampling hole. *(Blueray Systems, Inc.)*

PROCEDURE
1. Open the oil-supply line shut-off valve.
2. Set the room thermostat 5°F above room temperature.
3. After normal operating temperatures are reached, set draft 0.01 to 0.03 in (Figure 12–1). _____WC
4. Use smoke tester and set burner adjustment for maximum permissible amount of smoke (compare sample to chart).
5. Recheck draft and take CO_2 reading at sampling hole (Figure 12–1). _____% CO_2. If there is a large differential between CO_2 readings, air leakage is the most common cause. Reseal unit.
6. Open fire door, turn off oil valve, and check out safety timing of combustion control.
7. Check operation of limit controls and thermostat. (Disconnect belt or block off return air.)
8. Check for oil leaks.
9. Reset all controls for normal operation.

■ JOB SHEET 12–2 ■

Name _____

Score _____ Date _____

PERFORMANCE OBJECTIVE

Given an oil-fired furnace, make the inspections required for an installation follow-up, or a preseason tune-up.

REFERENCE

Heating, Ventilating, and Air Conditioning Fundamentals (Chapter 12)

EQUIPMENT

Oil-fired furnace

TOOLS

Hand tools, portable combustion test kit, wire brush with handle, vacuum cleaner

SUPPLIES

Shop towel, can of SAE 10 nondetergent oil

JOB 12–2

Dealing with preventive maintenance

PROCEDURE

1. Inspect the entire flue and venting system.
2. Replace the oil-filter cartridge in oil-supply line. (This should be done annually.)
3. Inspect nozzles for plugged distributor slots or plugged orifices. If necessary, replace the nozzle.
4. Lubricate the burner motor once every two years with a few drops of SAE 10 nondetergent oil.
5. Lubricate the supply fan motor and bearings (the sealed type requires no lubrication).
6. Inspect the belt (direct-drive blower units do not have belts).
7. Check the tubular portions of the heat exchanger internally for scale accumulation. Remove scale with a wire brush and a vacuum cleaner.
8. Replace the air filter with one of equal capacity. If the unit is supplied with a washable filter, it should be cleaned and coated with a special oil when dirty.
9. Check all wiring connections for damaged or frayed installation. Make sure all connections are tight.

■ JOB SHEET 12–3 ■

Name _____

Score _____ Date _____

PERFORMANCE OBJECTIVE
Given a domestic high-pressure gun-type burner, follow the normal servicing procedure outlined by the manufacturer for a specific burner. The job should not exceed the customer's estimate.

REFERENCE
Heating, Ventilating, and Air Conditioning Fundamentals (Chapter 12); Job 12–2

EQUIPMENT
High-pressure gun-type burner

JOB 12–3
Burner servicing

PROCEDURE
1. Caution: Turn off the power and oil supply.
2. Remove the combustion-head assembly. Clean the nozzle line and adaptor without wiping dirt over the nozzle orifice. Clean the electrode porcelains and the air-metering plate. Check the electrode settings. Set as shown in Figure 12–2.
3. Clean the inside of the blast tube. Remove any lint or dirt accumulation.
4. Clean the blower wheel. (Dirt and lint on the impeller blades reduce blower efficiency.)
5. Clean the combustion air openings around the air shutter and air band.
6. Reassemble the burner in the same manner you disassembled it. Install the air-metering plate and nozzle assembly into the blast tube. Make sure the nozzle specifications and air-metering plate specifications match the rating plate on the burner. The nozzle line should be loosely secured to the nozzle-line adjusting slide with a thumbnut. The assembly should be driven up tightly using the nozzle-line adjusting screw. The thumbnut and oil-line fitting should then be tightened. *The air-metering plate should be secure enough that no movement can be felt when applying pressure to the electrode buss bars.* Make sure the ignition transformer terminals make good contact with the buss bars. Secure the transformer to the housing with two locking screws.
7. Replace the in-line oil filter if necessary.
8. Open the oil supply and return power to the burner.
9. Check out the operation of the oil burner primary control and the fan limit control.
10. Follow the start-up adjustments before leaving the customer's house (Job 12–2).

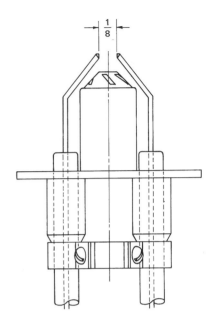

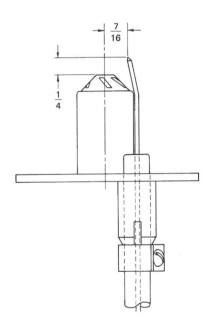

■ **FIGURE 12–2** Electrode settings.

■ MULTIPLE-CHOICE TEST ■

Name _____

Score _____ Date _____

DIRECTIONS
Circle the letter that best answers the following multiple-choice questions.

1. Number 2 grade fuel oil:

   a. will not burn
   b. is easily ignited
   c. has a flash point of 120°F (48.8°C)
   d. burns more readily than No. 1 grade oil

2. Manufacturers recommend 50 percent or less excess combustion air:

   a. mixed with the oil prior to ignition
   b. supplied by the burner
   c. regulated by the barometric damper
   d. induced by natural draft

3. Fuel oil is ignited by:

   a. an interrupted pilot
   b. a constant pilot
   c. an intermittent pilot
   d. a spark electrode

4. A low-pressure burner pushes the oil through the atomizing nozzle at 1 to 15 psi (6.8 to 103.4 kPa); oil pressure for a domestic high-pressure burner is adjusted to:

   a. 60 psi (413.6 kPa)
   b. 80 psi (551.5 kPa)
   c. 100 psi (689.4 kPa)
   d. 120 psi (827.3 kPa)

5. The burner motor drives a centrifugal fan that provides:

   a. primary air
   b. secondary air

   c. vaporization of the oil spray
   d. the proper oil pressure

6. The oil cut-off valve is:

   a. mounted within the burner draft tube
   b. part of the oil-pump pressure regulator
   c. commonly referred to as the nozzle
   d. nonadjustable

7. The first stage of a two-stage fuel unit:

   a. discharges to the tank when prime is established
   b. pumps directly to the bleed valve where air can be purged
   c. pumps oil to the nozzle bleed orifice
   d. opens the oil cut-off valve

8. The stack temperature should not exceed:

   a. 260°C
   b. 280 °C
   c. 300 °C
   d. 500 °C

9. The electrode transformer supplies:

   a. 100 mV
   b. 100 kV
   c. 0.10 kV
   d. 10 kV

10. The CO_2 reading of an efficient burner will be:

   a. 4 to 6 percent
   b. 6 to 8 percent
   c. 8 to 12 percent
   d. 12 to 15 percent

▓ WRITTEN EVALUATION ▓

1. How do you check the efficiency of a burner using a portable combustion test kit?
2. You've found a 1-in (25.4-mm) soot coating on the heat exchanger. What steps should you take to correct this problem?
3. The service complaint states that the unit is burning too much fuel. What would readily verify this complaint?

Unit 13

HEAT PUMP AND RECOVERY UNITS

■ INTRODUCTION

The following performance tasks require a visual inspection of the unit. The first thing an experienced service mechanic does on a service call is to look for obvious problems. You can do this by observing the various lines and components to find out if they are working properly. If the thermostat is calling for heating and the outdoor coil is blowing hot air, the first thing you should inspect is the reversing valve. Moreover, the solenoid coil on the reversing valve may be burned out so badly that you would not need a voltmeter to check it.

Before starting a heat pump, make the following checks:

1. Rotate the fan to check for frozen bearings or any binding.
2. Inspect all wiring for loose connections.
3. Make sure that all shut-off valves are properly positioned.
4. Check the unit serial plate for voltage requirements. And make sure that the voltage at the electrical disconnect switch is within 10 percent of the range listed on the unit.
5. Set the room thermostat for demand.
6. Recheck the voltage with the unit running.

Use the heat-pump diagram (Figure 13–1) when completing the first performance task.

FIGURE 13–1 Heat-pump diagram.

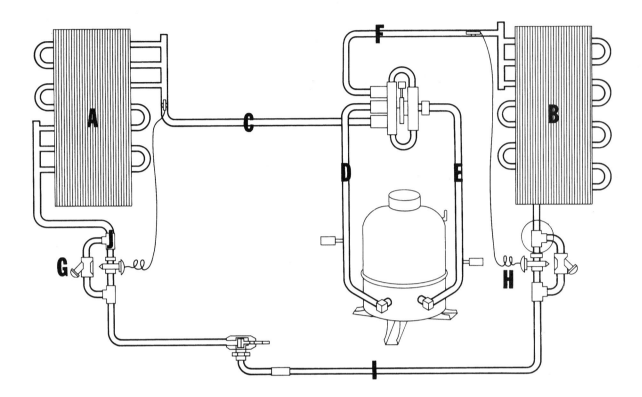

■ JOB SHEET 13–1 ■

Name _____

Score _____ Date _____

PERFORMANCE OBJECTIVE
Given a heat-pump diagram or a lab unit as shown in Figure 13–1, correctly identify the lettered components. State their function during the heating cycle.

REFERENCE
Heating, Ventilating, and Air Conditioning Fundamentals (Chapter 13); Figure 13–1

EQUIPMENT
Lab heat-pump optional

TOOLS
Hand tools

SUPPLIES
Notebook and pencil

JOB 13–1
Identifying component functions

PROCEDURE
1. Study the diagram in Figure 13–1. Start up the unit if it is available. Insert the correct letters in the appropriate blanks.
2. Set the thermostat to call for heating.
3. The low-pressure control is installed in the cold suction line _____.
4. The high-pressure cutout is located in the hot discharge line _____.
5. The outdoor coil *A* is being fed liquid by expansion valve _____.
6. A check valve is permitting liquid to bypass expansion valve _____.
7. Only during the heating cycle does line _____ function as the suction line.
8. Line _____ functions as the liquid line during the heating or cooling cycle.
9. Line _____ is functioning as the _____ line during the heating cycle.
10. Refrigerant is flowing through the drier *J* to *H* _____ or *H* to *J* _____. Check the correct answer.
11. Liquid flows through check valve *G* during the _____ cycle. Fill in the correct answer.

■ JOB SHEET 13–2 ■

Name _____

Score _____ Date _____

PERFORMANCE OBJECTIVE
Given a heat pump and air-pressure defrost control, make the outdoor coil ice up and test the performance of the defrost control system as described in text.

REFERENCE
Heating, Ventilating, and Air Conditioning Fundamentals (Chapter 13)

EQUIPMENT
Heat pump

TOOLS
Electronic temperature tester, volt-ohm-milliameter, hand tools

SUPPLIES
Piece of cardboard dimension of outdoor coil surface area, ice cubes, and water

JOB 13–2
Field-checking defrost control

PROCEDURE
1. Cut a piece of cardboard to cover approximately 85 percent of the outdoor coil surface area; remove condenser guard.
2. Mark position of defrost termination sensing bulb on refrigerant line; remove bulb from line and immerse in ice bath.
3. Start unit by adjusting room thermostat to call for heat.
4. Block coil inlet with cardboard from step 1. Unit should start defrost cycle.
5. After unit starts defrost cycle, remove sensing bulb from ice bath. Warm the bulb by holding in hand or immersing in warm-water bath. Unit defrost cycle should terminate.
6. After termination of defrost cycle, remount sensing bulb on refrigerant line in original location and secure clamp. Use caution to avoid damaging sensing bulb when installing.
7. If defrost control fails to operate properly, remove entire control and replace with a new control.

■ JOB SHEET 13–3 ■

PERFORMANCE OBJECTIVE
Given the proper tools and test instruments, estimate the air flow of a heat pump air handler. With stable air temperature, the cfm will be within plus or minus 10 percent.
Formula:

$$CFM = \frac{\text{watts } (3.413)}{\Delta t \, (1.08)}$$

REFERENCE
Heating, Ventilating, and Air Conditioning Fundamentals (Chapters 13 and 18)

TOOLS
Drill motor, ¼ in bit, pocket thermometer, VOM, amprobe (clamp-on ammeter)

MATERIAL
3 ¼-in plugs (to seal holes in duct)

JOB 13–3
Estimating air flow by temperature rise

PROCEDURE
1. Drill holes in supply air duct as shown in Figure 13–2. Holes must be within 6 ft of air handler and not in radiant heat area.

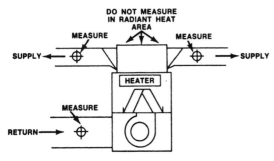

HEAT PUMP AIR HANDLER

■ **FIGURE 13–2** Estimating air flow (temperature rise method).

2. Measurement at supply and return registers will be inaccurate.
3. Average supply air readings.
4. Average supply air: _____°F; Return air: _____°F
5. Delta t (T.R.) _____ °F
6. Find resistive load (watts) = volts _____ × amps _____

7. Watts = _____

8. $\text{Cfm} = \dfrac{\underline{\hspace{1cm}} \text{ (watts)} \times 3.413 \text{ (BTU/watt)}}{\underline{\hspace{1cm}} \text{ T.R.} \times 1.08 \text{ (constant)}}$

9. Cfm = _____

■ JOB SHEET 13–4 ■

Name _____

Score _____ Date _____

PERFORMANCE OBJECTIVE
Given the proper guidelines, check, test, and start a water-to-water heat pump. Unit will perform as designed or a list of corrections needed will be itemized.

REFERENCE
Heating, Ventilating, and Air Conditioning Fundamentals (Chapter 13); Florida Heat Pump installation manual

TOOLS
Tape measure, refrigeration gauges, thermometers, amprobe, VOM, water gauges, assorted screwdrivers

SUPPLIES
Notebook and pencil

JOB 13–4
Start up a water-to-water heat pump

PROCEDURE
1. Deenergize unit.
2. Check that high-voltage power supply is correct and in accordance with serial plate. _____
3. Check that field wire and circuit protection are correct size (refer to Chapter 9 text).
   Wire size: _____AWG
   Circuit breaker: _____amps; Fuses: _____amps
4. Low voltage transformer:
   Primary: _____208 V or _____230 V
   Secondary: _____volts
5. Check that piping system is complete and correct. (Refer to Figures 13–3 and 13–4.)
6. Provide vibration isolation.
7. Check that unit is serviceable.
8. Replace unit access panels.
9. Turn thermostat to the "off" position.

INITIAL UNIT START-UP
1. Set thermostat to highest position.
2. Set system switch to cool. Compressor should not run.
3. Reduce the thermostat setting until the compressor, reversing valve, and loop pump are energized. Adjust water flow utilizing pressure temperature plugs and specification sheet data furnished with unit.
4. Check the cooling refrigerant pressures against the values in Table 13–1.

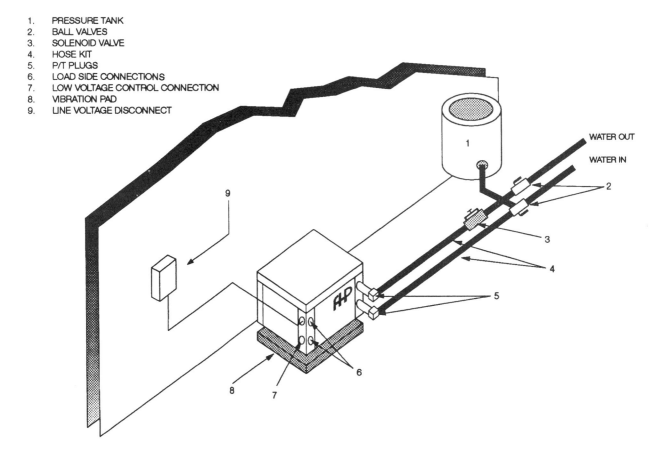

1.   PRESSURE TANK
2.   BALL VALVES
3.   SOLENOID VALVE
4.   HOSE KIT
5.   P/T PLUGS
6.   LOAD SIDE CONNECTIONS
7.   LOW VOLTAGE CONTROL CONNECTION
8.   VIBRATION PAD
9.   LINE VOLTAGE DISCONNECT

FIGURE 13–3   Well water application (source side).

5. Turn the system switch to "off." Unit should stop running and reversing valve should deenergize.
6. Leave unit off for five minutes to allow for pressure equalization.
7. Turn thermostat to lowest setting.
8. Set thermostat system switch to "Heat Position."
9. Adjust temperature setting upward until unit is energized.
10. Check the heating refrigerant pressures against values in Table 13–2.
11. Check for vibration, leaks, etc.
12. Set thermostat to maintain desired space temperature.
13. Instruct owner on system operation.

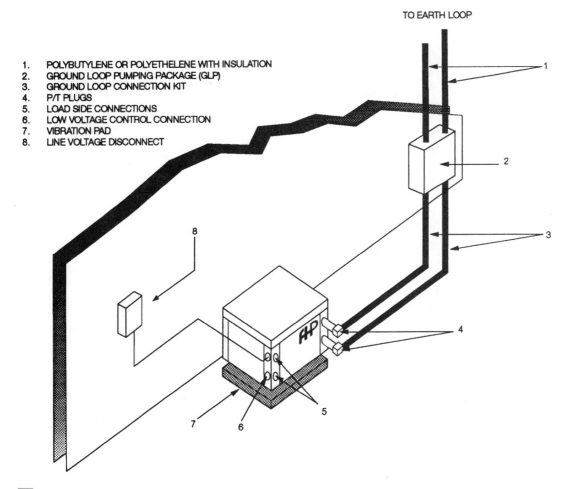

TO EARTH LOOP

1. POLYBUTYLENE OR POLYETHELENE WITH INSULATION
2. GROUND LOOP PUMPING PACKAGE (GLP)
3. GROUND LOOP CONNECTION KIT
4. P/T PLUGS
5. LOAD SIDE CONNECTIONS
6. LOW VOLTAGE CONTROL CONNECTION
7. VIBRATION PAD
8. LINE VOLTAGE DISCONNECT

FIGURE 13–4   Earth-coupled application (source side).

TABLE 13–1   Cooling mode operating pressures.

ENTERING LOAD DEG F.	ENTERING SOURCE DEG F.	SUCTION PRESSURE	DISCHARGE PRESSURE
60	30	40-50	160-190
	60	65-75	220-250
	90	85-95	260-290
90	30	45-55	190-220
	60	70-80	255-285
	90	90-100	295-325
120	30	50-60	220-250
	60	75-85	280-320
	90	95-105	320-350

TABLE 13–2   Heating mode operating pressures.

ENTERING LOAD DEG F.	ENTERING SOURCE TEMPERATURE DEG F					
	55		75		95	
	SUCT	DISCH	SUCT	DISCH	SUCT	DISCH
60	60-65	120-160	65-70	15-180	70-75	195-235
90	72-77	150-190	77-81	175-215	80-85	225-265
120	83-88	180-270	86-91	205-245	90-95	250-290

■ JOB SHEET 13–5 ■

Name _____

Score _____ Date _____

PERFORMANCE OBJECTIVE
Given a water-to-water heat pump that is not working, properly go through an orderly troubleshooting procedure. Evaluation should take less than one hour.

REFERENCE
Heating, Ventilating, and Air Conditioning Fundamentals (Chapter 13); Florida Heat Pump installation manual

TOOLS
Tape measure, refrigeration gauges, thermometers, amprobe, VOM, water gauges, assorted screwdrivers

SUPPLIES
Notebook and pencil

JOB 13–5
Troubleshoot a water-to-water heat pump (Figure 13–7)

PROCEDURE

COMPRESSOR DOES NOT OPERATE
1. Check power supply for adequate voltage.
2. Check control circuit for proper connections. (See Figure 13–5, Typical Wiring.)
3. Check for loose wires.
4. Check run capacitor (textbook Chapter 3).
5. Check internal compressor ground or open winding (procedure outlined in textbook Chapter 3).

OPEN HIGH-PRESSURE SWITCH
1. Check condenser water flow: 3 gpm/ton with cooling tower; 1½ gpm/ton city water.
2. Check water temperature:
   Approximately 80°F entering; 90°F leaving (cooling tower, 10 degree rise; city water, 20 degree rise).
3. Check for scaled or fouled condenser:
   High pressure drop, low temperature rise, high head.
4. Check for air in system (use pressure/temperature chart).

OPEN LOW-PRESSURE SWITCH
1. Check water flow (see Figure 13–6).
2. Check water temperature.
3. Check for loss of refrigerant.

INSUFFICIENT COOLING OR HEATING
1. Check thermostat for proper location. Avoid outside walls and drafts.
2. Check water flow (see Figure 13–6).
3. Check refrigerant charge.
4. Check system components.

2 Hermetic compressors, (WP-reversing valve), expansion valve for refrigerant metering. High and low refrigerant pressure switches and a lock-out impedance relay. Pump relay, crankcase heaters, schrader tap for water regulating valves. PTCR on all single phase units, low pressure time delay relay, power block, and low voltage terminal strip. Finished in beige.

WP – HEAT PUMP
WH – HEATING ONLY

 **SPECIFICATION DATA SHEET**

WP/WH SERIES
MODELS 124, 156

WATER TO WATER MODELS

CAPACITY DATA – WATER TO WATER RATINGS

MODEL ▶	WH/WP 124	WH/WP 156
COOLING BTUH	130013	149964
HEATING BTUH	128004	158955
LOAD GPM	24.9	30.8
SOURCE GPM	30	36

RATING CONDITIONS: ENTERING SOURCE LIQUID TEMP: 55°F.
LOAD SOURCE LIQUID TEMP: HEATING: 120°F. ENTERING. 130°F. LEAVING.
COOLING: 45°F. LEAVING.

ELECTRICAL DATA

MODEL WH/WP	POWER SUPPLY	COMPRESSOR RLA/LRA (EA.)	TOTAL FLA *	MIN. CIR. AMPACITY *	MAXIMUM FUSE *
124-3	208-230/3/60	15.7 / 110.0	55.4	59.3	75
124-4	460/3/60	7.8 / 55.0	27.6	29.6	35
124-5	575/3/60	6.0 / 30.0	20.0	21.5	25
156-3	208-230/3/60	21.6 / 135.0	67.2	72.6	90
156-4	460/3/60	10.6 / 70.0	33.2	35.9	45
156-5	575/3/60	8.2 / 39.0	24.4	26.5	30

* – INCLUDES 24 AMPS FOR WATER PUMP AT 230V. INCLUDES 12 AMPS FOR WATER PUMP AT 460V, 8 AMPS AT 575V.

MECHANICAL SPECIFICATIONS

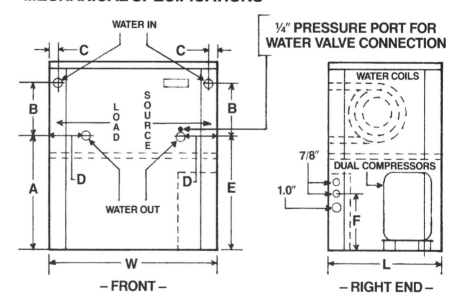
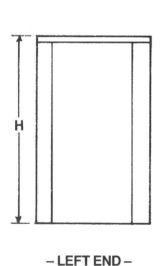

– LEFT END –   – FRONT –   – RIGHT END –

FIGURE 13–5 Typical wiring for water-to-water heat pumps.

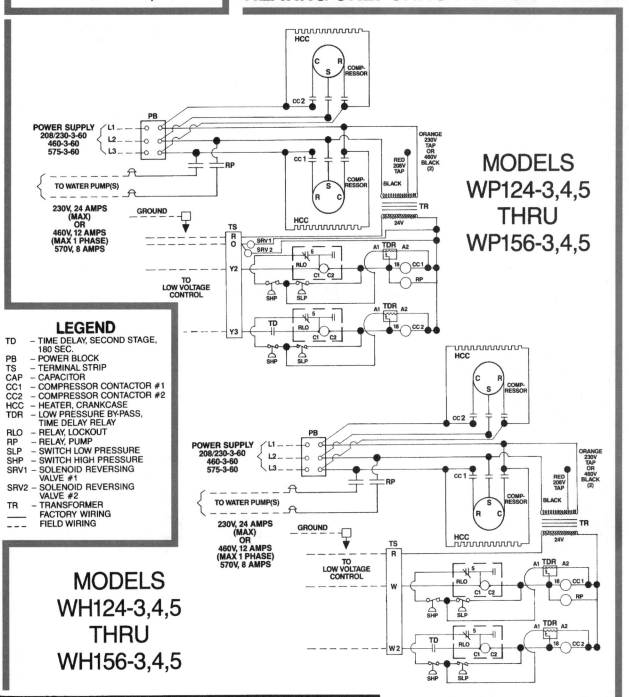

WP/WH SERIES
MODELS 124, 156

FHP SPECIFICATION DATA SHEET
**WATER TO WATER HEAT PUMPS,
HEATING ONLY UNITS – TYPICAL WIRING**

MODELS
WP124-3,4,5
THRU
WP156-3,4,5

LEGEND

TD – TIME DELAY, SECOND STAGE, 180 SEC.
PB – POWER BLOCK
TS – TERMINAL STRIP
CAP – CAPACITOR
CC1 – COMPRESSOR CONTACTOR #1
CC2 – COMPRESSOR CONTACTOR #2
HCC – HEATER, CRANKCASE
TDR – LOW PRESSURE BY-PASS, TIME DELAY RELAY
RLO – RELAY, LOCKOUT
RP – RELAY, PUMP
SLP – SWITCH LOW PRESSURE
SHP – SWITCH HIGH PRESSURE
SRV1 – SOLENOID REVERSING VALVE #1
SRV2 – SOLENOID REVERSING VALVE #2
TR – TRANSFORMER
—— FACTORY WIRING
--- FIELD WIRING

MODELS
WH124-3,4,5
THRU
WH156-3,4,5

FHP MANUFACTURING
601 N.W. 65TH COURT
FT. LAUDERDALE, FL 33309
PHONE: (305) 776-5471
FAX: 776-5529

As a result of continuing research and development, all ratings and specifications are subject to change without notice.

a ◼HARROW company

FIGURE 13–6 Specification data sheet—water-to-water heat pump.

OPEN HIGH PRESSURE SWITCH.
2. a. Check water flow.
    b. Check water temperature.
    c. Check for scaled or fouled condenser.

OPEN LOW PRESSURE SWITCH.
3. a. Check water flow.
    b. Check water temperature.
    c. Check for scaled or fouled condenser.
    d. Check for loss of refrigerant charge

* **NOTE**: If lock-out on safety circuit occurs check load side heat transfer surface for adequate size.

INSUFICIENT COOLING OR HEATING:
1. Check thermostat for proper location. Avoid outside walls and drafts.
2. Check water flow.
3. Check refrigerant charge.
4. Check system components.

OPTIONS:

Heat recovery package. A factory installed heat recovery package is available for potable water heating . See HRP literature.

Hot gas bypass – A factory installed hot gas bypass option is available for capacity control.

Ground loop pumping package – is available for field installation. See GLP literature.

Control Options – Various control options are available such as time delay relays, random start relays, aquastats, etc. Consult factory for application assistance.

HYDRONIC IN-SLAB FLOOR HEATING

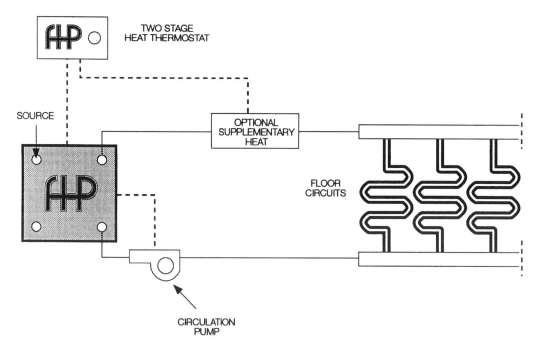

FIGURE 13–7   Hydronic in-slab floor heating with chilled-water cooling.

■ JOB SHEET 13–6 ■

Name _____

Score _____ Date _____

PERFORMANCE OBJECTIVE
Select a heat pump and a matching heat recovery unit for a hydronic in-slab floor heating and chilled-water cooling system. Specifications will match load calculations.

REFERENCE
Heating, Ventilating, and Air Conditioning Fundamentals (chapters 7 and 13); Florida Heat Pump specification data sheet; and Doucette heat recovery unit tables

TOOLS
Notebook and pencil

SUPPLIES
Manufacturer's specifications and data sheet, pictorial drawing of installation

JOB 13–6
Select a water-to-water heat pump and a heat recovery unit.

PROCEDURE
1. Record heating load (Chapter 7, Figure 7–5): _____Btu/h
2. Record cooling load (Figure 7–4): _____Btu/h
3. Heat pump model #(refer to Figure 13–6): _____
4. Circuit ampacity: _____; Fuse size: _____
5. Heat recovery unit model #: _____(Table 13–3a).
6. Compressor runs 7 hour/day (cooling season). Domestic hot water recovery: _____gallons/day

TABLE 13–3 Selection and performance of heat recovery unit.

Model	System Tons	60° F to 100° F BTUH Recovery GPM		60° F to 140° F BTUH Recovery GPM		100° F to 140° F BTUH Recovery GPM	
SD 5	1.0	3,196	0.16	2,641	0.07	2,200	0.11
SD 10	2.0	6,391	0.32	5,282	0.13	3,962	0.20
SD 15	3.0	9,680	0.48	8,000	0.20	6,000	0.30
SD 20	4.0	13,310	0.67	11,000	0.28	8,250	0.41
SD 20	5.0	14,520	0.73	12,000	0.30	9,000	0.45
SD 30	7.5	21,780	1.09	18,000	0.45	13,500	0.68
SD 40	10.0	28,992	1.45	24,000	0.60	19,920	1.00
SD 60	15.0	46,188	2.31	36,000	0.90	29,880	1.49
SD 80	20.0	62,867	3.14	49,000	1.23	40,670	2.03
SD 120	25.0	85,008	4.25	69,000	1.73	57,960	2.90

Model	System Tons	60° F to 100° F BTUH Recovery GPM		60° F to 140° F BTUH Recovery GPM		100° F to 140° F BTUH Recovery GPM	
SD 5	1.0	2,490	0.12	1,500	0.04	1,088	0.05
SD 10	2.0	5,518	0.28	3,324	0.08	2,410	0.12
SD 15	3.0	8,277	0.41	4,986	0.12	3,615	0.18
SD 20	4.0	11,786	0.59	7,100	0.18	5,148	0.26
SD 20	5.0	13,795	0.69	8,310	0.21	6,025	0.30
SD 30	7.5	20,692	1.03	12,465	0.31	9,037	0.45
SD 40	10.0	27,589	1.38	16,620	0.42	12,050	0.60
SD 60	15.0	36,520	1.83	22,000	0.55	15,950	0.80
SD 80	20.0	55,178	2.76	33,240	0.83	24,099	1.20
SD 120	25.0	68,973	3.45	41,550	1.04	30.124	1.51

Water Flow (GPM/Ton)	0.1 Min.	0.2	0.3	0.4	0.5 Max.
PSID	0.6	1.8	3.7	6.2	9.0

■ MULTIPLE-CHOICE TEST ■

Name _____

Score _____ Date _____

DIRECTIONS
Circle the letter that best answers the following multiple-choice questions.

1. An air-to-air heat pump:

    a. has an air-cooled condenser
    b. has a water-cooled condenser
    c. has an evaporative condenser
    d. picks up heat from air and releases heat to air

2. A heat pump compared to fossil fuel furnaces:

    a. has a longer running time
    b. delivers the same amount of heat
    c. maintains a more uniform temperature
    d. all of the above

3. During the cooling cycle of a heat pump:

    a. latent heat of vaporization takes place in the outdoor coil
    b. latent heat of vaporization takes place in the indoor coil
    c. latent heat of condensation takes place in the indoor coil
    d. both *a* and *c* are correct

4. The reversing valve of a heat pump is called a four-way valve because it connects four different refrigerant lines. Its pilot-control valve has:

    a. one line
    b. two lines
    c. three lines
    d. four lines

5. Air-to-air heat pumps are the most commonly found because they are:

    a. more efficient
    b. less expensive
    c. easier to service
    d. easy with respect to the reverse cycle

6. A body of air or liquid to which heat removed from the home is transferred is called:

    a. the indoor coil
    b. the outdoor coil
    c. the heat sink
    d. the reverse-cycle pond

7. The heating efficiency of a heat pump is indicated by:

    a. EER rating
    b. COP rating
    c. sensible heat ratio
    d. its dehumidification capability

8. The coefficient of performance rating for an electric furnace is generally:

    a. equal to that of a heat pump
    b. greater than that of a heat pump
    c. 3.41 Btu/W or 1.0 Btu/W
    d. the same as its power factor

9. Supplementary heat is needed:

    a. at low ambient conditions
    b. when the thermostat calls for two-stage heat
    c. at the outdoor-temperature balancing point
    d. when the room temperature is below the thermostat set point

10. When comparing a heat pump to a gas-fired or oil-fired furnace, a true measure of the efficiency is:

    a. EER
    b. COP
    c. SPF
    d. Btu

WRITTEN EVALUATION

1. State the advantage of using a pilot-operated reversing valve.
2. Compare heat-pump components to those found on a straight cooling unit.
3. Give the reasons for using supplementary heat and the existing conditions when it is applied.

Unit 14
SOLAR SYSTEMS

■ INTRODUCTION

The following performance tasks involve the design, installation, start-up, and testing of various solar-heating systems. Field installations may vary widely in design, operation, and performance due to climatic conditions and component compatibility. The following tasks, however, will provide a general understanding of how solar systems collect and convert solar radiation into usable energy.

■ COLLECTOR ORIENTATION

For optimum performance the collector should be oriented true south. Do not confuse true south with magnetic south that is provided by a compass reading. The difference between magnetic south and true south is called the *magnetic declination adjustment.*

To find the declination angle for your area ask a local surveyor or look at an isogonic map of the United States. You can also obtain a plot map from your local tax office. Examples of declination angles are: Los Angeles, California, 16° west of magnetic south; Miami, Florida, 0° of magnetic south; Pittsburgh, Pennsylvania, 6° east of magnetic south. Maine, the northeastern tip of the United States, has a declination angle of 22° east of magnetic south and northwestern boundary, Seattle,

Washington, is 22° west of magnetic south. If collectors were oriented magnetic south from these extremities, there would be a loss in efficiency of approximately 5 percent. Therefore, slight alterations from true south will not result in a great heat loss and sometimes a collector oriented 5 to 10° west of true south will pick up more heat.

■ COLLECTOR TILT

The National Solar Heating and Cooling Information Center gives the following rules of thumb for flat-plate collector tilt angles:

1. For a system used only for domestic hot water, the tilt angle is equal to latitude. [See Appendix weather data for latitude of your area (Figure A–3).]
2. For systems used only for space heating, the collector tilt angle should be latitude plus 15°.
3. For systems used for domestic hot water and heating, the collectors should be tilted latitude plus 15°.
4. A collector used for heating and cooling should be tilted latitude plus 5°.
5. Swimming-pool collectors should be tilted at an angle equal to latitude for year-round application; or latitude minus 10 to 15° for summertime-only applications.

▦ MOUNTING COLLECTORS

In addition to having hydronic and air collectors, consideration must be given to whether the collectors are to be mounted on a new roof, an existing roof, or a remote installation. Figure 14–1 illustrates a collector installation between the rafters, which is appropriate for a new roof.

Installation of air collectors appropriate for an existing roof is depicted in Figure 14–2.

A remote installation is shown in Figure 14–3. The water lines or air ducts are run underground to the residence.

Step-by-step installation instructions for hydronic collectors on a roof-deck mount follow:

1. Roof deck should be papered with desired felt.
2. Supportive 2 × 4 should be laid out, squared, and leveled off roof edge. Remaining 2 × 6 flashing framework can be built at the same time.
3. Set first panel on deck starting at desired end. Unit can be set by crane or by hand. If set by hand, edges or felt should be secured to prevent catching on collector edge.
4. Secure panel by angle clips (provided).
   a. Clips should be a maximum of 48-in OC.
   b. Attach clip to panel using drill and sheet-metal screws provided.

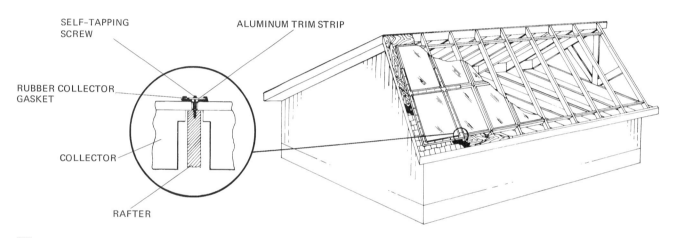

▦ **FIGURE 14–1** Installation between rafters. *(Research Products Corp.)*

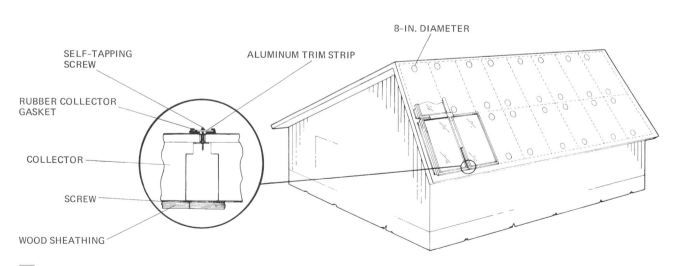

▦ **FIGURE 14–2** Installation on roof sheathing. *(Research Products Corp.)*

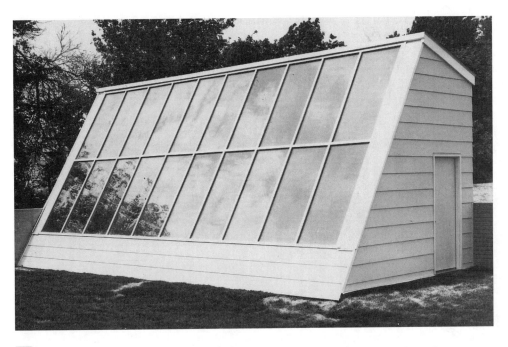

■ FIGURE 14–3 Remote installation. *(Research Products Corp.)*

   c. Lag bolt clip to roof (should align with truss).

5. Set panel 2 in place in accordance with steps 3 and 4.

6. Place cap strip retainer (supplied) in continuous slot between panels. (Strips slide in from bottom; turn one corner to prevent strip from sliding out.)

7. Place cap strips over panels 1 and 2; retain with screws 24-in OC. (Strips should not be tightened to extremes that could break glass.)

8. Repeat steps 3 through 7 until all panels, retainers, and cap strips are in place.

9. Collector return and supply headers should enter and exit from top and bottom of collector, respectively. Each header will sweat-fit to a T connection, and supply and return lines can be piped accordingly between collectors. (See hydronic piping details.)

10. Complete T attachment, return, and supply lines to all panels.

11. Measure for flashing (supplied by others) and install according to factory-recommended flashing details.

■ PIPING DETAILS

The isometric drawing of a self-draining hot-water system shown in Figure 14–4 details the piping and components of an open system. The collectors drain when the storage tank circulator pump is turned off. The drain-down open system is for freeze protection.

A closed-system piping is shown in Figure 14–5. It employs a water-storage circulating pump and a glycol (antifreeze solution) pump.

Additionally, closed systems require water makeup, a compression tank, and relief valves (Figure 14–6).

Avoid these errors when installing the B&G Airtrol system:

1. Tank hung insecurely. Nails too small and loosened when water enters tank, causing—

2. Pitch of air line in wrong direction. Using pipe strap which has been folded will also cause tank to sag with added weight.

3. Square head cock in line. Never install a valve in the horizontal part of air line. If a valve must be used, install a gate valve in the vertical pipe line.

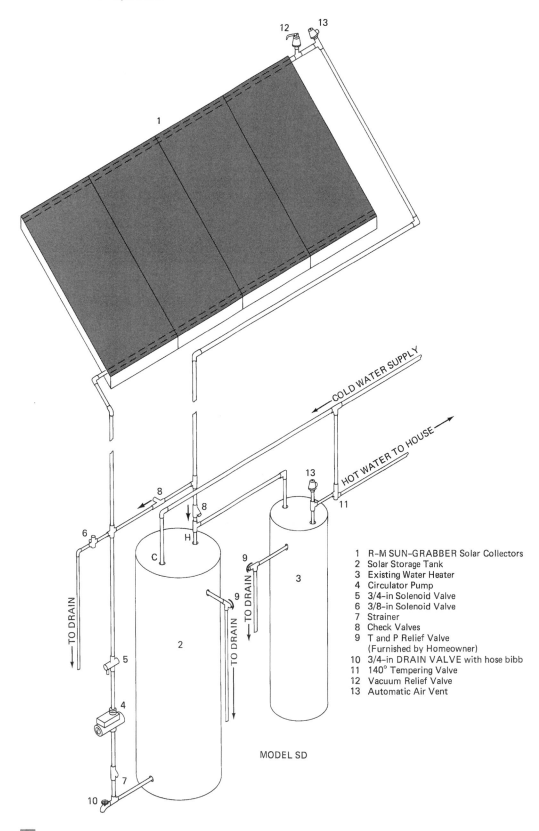

1 R-M SUN-GRABBER Solar Collectors
2 Solar Storage Tank
3 Existing Water Heater
4 Circulator Pump
5 3/4-in Solenoid Valve
6 3/8-in Solenoid Valve
7 Strainer
8 Check Valves
9 T and P Relief Valve
   (Furnished by Homeowner)
10 3/4-in DRAIN VALVE with hose bibb
11 140° Tempering Valve
12 Vacuum Relief Valve
13 Automatic Air Vent

MODEL SD

FIGURE 14–4   Self-draining domestic hot-water system. *(R-M Products.)*

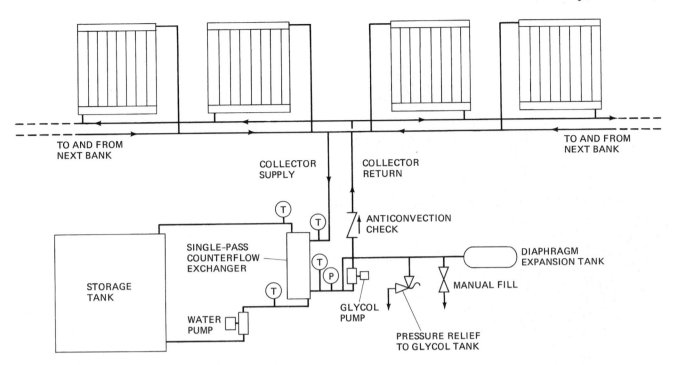

■ **FIGURE 14–5**   Closed-system collector piping. *(R-M Products.)*

4. Union should be in the vertical instead of horizontal section of pipe line.
5. Bushings in both Airtrol tank and boiler fitting. Pipe is too small to allow free passage of air in opposite directions.
6. Relief valve installed in air line. This can result in loss of air from tank if relief valve operates.
7. Nipple between Airtrol boiler fitting and the boiler too long. It must be as short as possible. Tube must be submerged in boiler water or there is no effective air trap. After boiler fitting is made up, push adjustable tube down as far as possible.
8. Frequent venting of Airtrol tank fitting will result in a water-logged tank. This vent should be opened to release air trapped in the tank only when the system is first filled and placed in operation.
9. Do not use the tapping in the Flo-Control valve for the connection to the compression tank. Use the tapping on the Airtrol boiler fitting.

The piping for a thermosyphon system is shown in Figures 14–7a and 14–7b. This is practical for warm-climate areas where freeze protection is not needed. The cold water enters the lower collector manifold. The storage tank must be at least 2 ft above the collectors.

■ POOL HEATERS

Glazing is not needed for pool collectors. (See Figure 14–8.) Moreover, collectors can be installed in many ways: *(a)* almost entirely out of view; *(b)* on the third-floor roof of an apartment complex; *(c)* on a frame or bank adjacent to the pool; or *(d)* on the roof of a garage. The last part *(e)* shows the controls.

Figures 14–9a and 14–9b display an automatic control system's components. The solar sensor turns the system on and the control valve bypasses the collectors at a set temperature to prevent overheating the water.

■ PASSIVE SYSTEM

In Japan, there are thousands of solar water heaters being used. Many are nothing more than a black box on a roof filled with 30 gallons (113.5 L) of water. In the morning the tank is filled with tap water and by late afternoon the tank is heated to over 100°F (37.7°C).

The average American family uses approximately 20 gallons (75.7 L) of water per person. Thus, a family of four would need a storage tank of 80-gal capacity (302.8 L) and from 40 to 60 ft^2 (3.7 to 5.5 m^2) of solar collectors to meet their needs.

INSTALLATION ON TOP OUTLET BOILERS

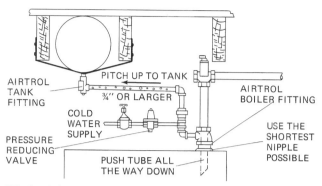

FIG. 1. Horizontal piping between boiler and compression tank must be full size of tapping in the Airtrol Tank Fitting. If horizontal pipe length is more than 7 feet, increase to next larger size pipe—two sizes larger if horizontal pipe is more than 20 feet. *Do not use a valve of any kind between the compression tank and boiler!* It is unnecessary and prevents free passage of air in the tank. If a valve must be used, install a gate valve in the vertical pipe line.

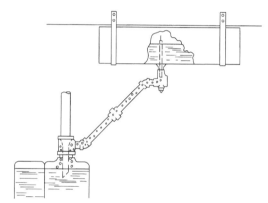

FIG. 2. This is an ideal method of running the pipe between the boiler and compression tank, as it permits an unrestricted flow of air bubbles to the tank. When this type of connection is not practical, horizontal piping with sufficient pitch-up to the tank (see Fig. 1) is adequate. A minimum of 1 in. pitch-up in 5 feet should be used.

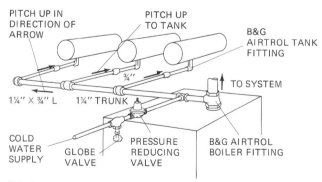

FIG. 3. Where there is not sufficient space between the boiler and the ceiling for a single compression tank of adequate capacity, several smaller tanks may be used. When two tanks are used, increase the horizontal header to one size larger than the tapping in the Airtrol Tank Fitting. For three or more tanks in parallel, increase the header two sizes. In installations where ceiling height will not permit unions in vertical piping they may be used horizontally. Airline piping must pitch-up to tanks.

INSTALLATION ON SIDE OUTLET BOILERS

Side or end outlet boilers are generally tapped for forced circulation pipe sizes. Therefore ABFSO Airtrol Fittings are furnished with the same size supply and boiler connections. If it is necessary to use a different main size, reduce or increase at the *system* tapping, never at the boiler connection. The dip tubes on ABFSO Fittings are not adjustable. The Fittings must there-fore be connected with a close or shoulder nipple, so that the dip tube will extend into the boiler.

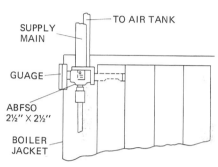

FIG. 4. In some side outlet boilers, the ABFSO Boiler Fitting must be installed inside the jacket. In this case, after the Fitting is installed, a 1 1/8-in hole should be cut in the top of the jacket for the 3/4-in pipe connection to the compression tank. Do not bush the 3/4-in tapping. The ABFSO-2 1/2 x 2 1/2 size is furnished with an additional 3/4-in tapping at the bottom which may be used for a mechanical type gas control.

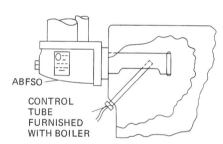

FIG. 5. Another type of side outlet boiler has a mechanical control which is screwed into the front of the boiler, below the outlet tapping and projects into the nipple port passage. In boilers of this type, the ABFSO Fitting must be installed before the tube of the mechanical control is inserted.

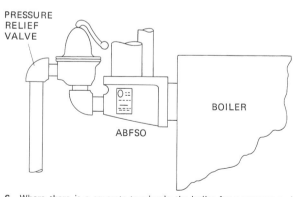

FIG. 6. Where there is a separate tapping in the boiler for a pressure and temperature gauge, the tapping on the end of the ABFSO Fitting may be used for the pressure relief valve. Do not install the relief valve in the line between the ABFSO Fitting and the compression tank.

FIGURE 14–6 Airtrol tank fittings. *(ITT-Bell and Gossett Co.)*

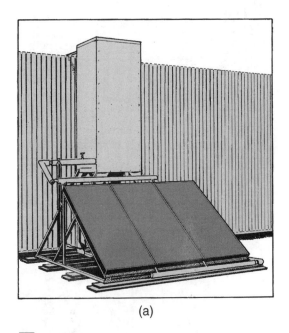

Needs vary from one neighborhood to another, as well as from one country to another. Likewise, the following solar projects permit the student to build a black box or a more elaborate system. Either a small-scale project or a large-scale project can be used to demonstrate how a passive system performs.

(a)

FIGURE 14–7a   Thermosyphon system. *(Raypak, Inc.)*

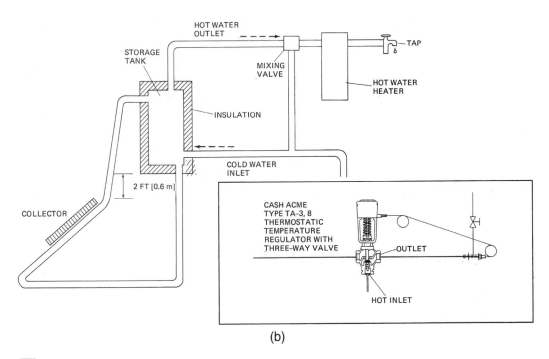

(b)

FIGURE 14–7b   Detailed thermosyphon system.

(a)

(d)

(b)

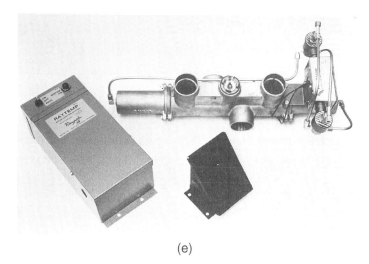

(e)

(c)

FIGURE 14–8 Solar pool collectors. *(Raypak, Inc.)*

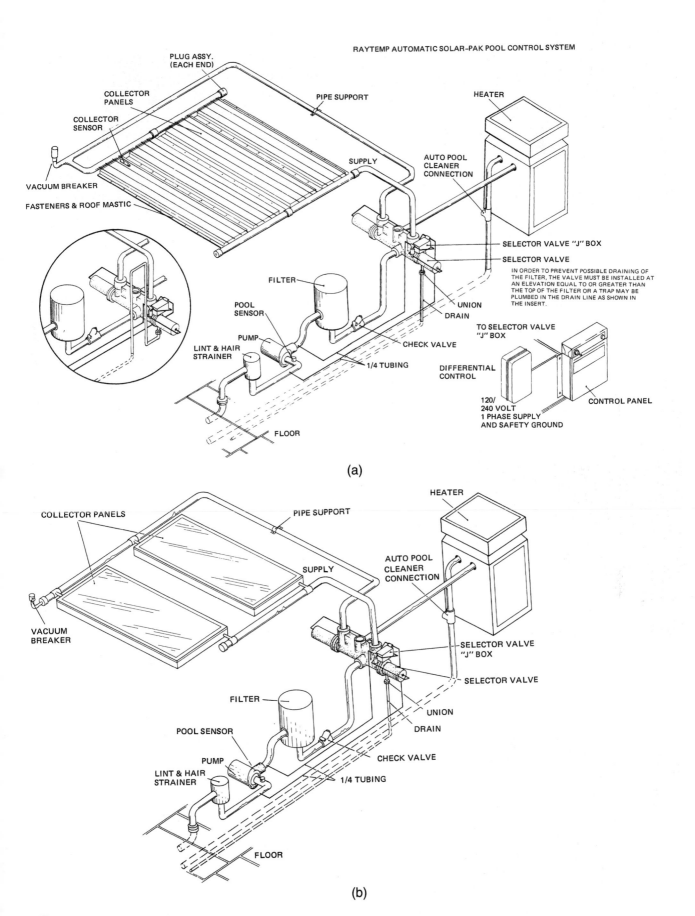

(a)

(b)

FIGURE 14–9 System diagrams: *(a)* automatic and *(b)* manual control. *(Raypak, Inc.)*

■ JOB SHEET 14–1 ■

PERFORMANCE OBJECTIVE
Given a boiler, compression tank, Airtrol tank fitting, and closed hydronic system piped and ready for filling, properly fill a system to maintain an air cushion. Time limit will vary with system size and water pressure available.

REFERENCE
Heating, Ventilating, and Air Conditioning Fundamentals (Chapter 14); Figure 14–6

EQUIPMENT
Hydronic system with boiler

TOOLS
Hand tools

JOB 14–1
Filling a closed hydronic system

PROCEDURE
1. Close all air vents, except the vent at the bottom of the Airtrol valve fitting (see Figure 14–6) and begin filling system. Leave the ATF vent open until water runs freely from it; then, close tightly. Do not open this vent again unless system has to be drained.
2. Vent radiation units and high points of the system.
3. Adjust the B&G pressure-reducing valve to provide pressure to the highest point of the system.
4. After the system has been completely filled, start the pump and allow it to circulate cold water for a short time. This will dislodge entrained air bubbles in the system and return them to the air separator.
5. Stop the pump and start the boiler burner. Allow the boiler temperature to reach 220°F (104.4°C), then stop firing. Wait at least a minute, then start the pump. (If any radiation is of the panel type, which may be damaged by extremely hot water, allow the boiler to cool to 140°F [60°C] before starting the pump.)
6. Stop the pump. Vent radiation and high points of the system. Normal operation may now be started at any time.

■ JOB SHEET 14–2 ■

Name _____

Score _____ Date _____

PERFORMANCE OBJECTIVE
Given solar collector(s), solar rack kit, piping, and storage tank, properly orient and tilt collectors. Maximum temperatures should be reached at prerecorded orientation data.

REFERENCE
Heating, Ventilating, and Air Conditioning Fundamentals (Chapter 14); Figure 14–10; weather data information (Figure A–3)

EQUIPMENT
Open or closed hydronic solar system with collectors mounted on an adjustable rack

TOOLS
Level, protractor, hand tools, socket set, temperature-test equipment, compass

JOB 14–2
Properly orienting and tilting collectors

PROCEDURE
1. Determine latitude from weather data (Figure A–3).
2. Use compass to face collectors magnetic south.
3. Increase tilt angle 15° over latitude (step 1). Measure with protractor.

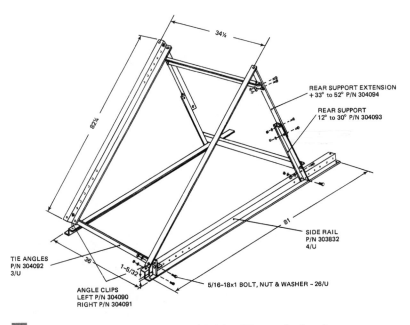

FIGURE 14–10 Solar rack kit. *(Raypak, Inc.)*

4. Fill system as outlined in Job 14–1.
5. Turn on the working fluid pump.
6. Record temperature readings at differential controller sensors.
7. Mark with chalk collector location rack.
8. Slowly reorient the collector test rack while checking thermometer recordings. Prepare to fill in the blanks below.
9. Optimum results were obtained at:

    _____compass reading

    _____angle of tilt
10. Record the declination angle of the collectors (difference in degrees between magnetic south and true south): _____
11. Check a topographic map of your area or an isogonic map of the United States to find the declination angle. Does it differ from step 10?

■ JOB SHEET 14–3 ■

Name _____

Score _____ Date _____

PERFORMANCE OBJECTIVE
Given a hydronic heating system, calculate the equivalent boiler horsepower rating (Bhp).

REFERENCE
Heating, Ventilating, and Air Conditioning Fundamentals (Chapter 14)

EQUIPMENT
Hot-water heating system with an array of solar hydronic collectors (install an in-line flow meter if not provided with the system), electronic temperature gauge, tape measure, notebook, calculator

JOB 14–3
Determining equivalent boiler horsepower rating of an array of flat-plate solar collectors and finding the insolation factor (Btu/h $\times$ ft^2) of the solar collectors

PROCEDURE
1. With an electronic thermometer, measure the water inlet temperature of the collectors. _____°F; _____°C
2. Measure collector outlet water temperature. _____°F; _____°C
3. Measure the collector area. _____m^2; _____ft^2
4. The working fluid pump should deliver from 0.025 gal/(min)(ft)^2 to 0.05 gal/(min)(ft)^2 of collector. What does your in-line flow meter indicate? _____gal/min
5. Multiply gal/min (step 4) $\times$ 0.000063. _____m^3/s
6. Multiply gal/min (step 4) $\times$ 0.063. _____L/s
7. The Btu/h is equal to 500 $\times$ gal/min $\times$ the water temperature rise. From step 4 determine _____Btu/h
8. Pumps are selected to deliver 0.025 to 0.05 gal/(min)(ft)^2 of collector. For 80 ft^2 of collector and a flow rate of 0.025, calculate the Btu/h using the temperature difference found (steps 1 and 2). _____Btu/h
9. For 80 ft^2 of collector and a flow rate of 0.05, with the temperature rise found in steps 1 and 2, find _____Btu/h
10. Does the collector pump on your installation fall in the range of steps 8 and 9? _____
11. What is the insolation factor [Btu(h)(ft)^2] of the collector array? _____Btu(h)(ft^2)
12. The equivalent boiler horsepower (Bhp) for a solar hot-water system is 33,475 Btu/h. What is the Bhp of your solar collector array?
    500 $\times$ gal/min $\times$ collector temperature rise = _____Btu/h
    Btu/h $\div$ 33,475 _____Bhp

13. Bhp (step 12) $\times$ 9,809.5 = _____W
14. W (step 13) $\times$ 3.412 = _____Btu/h
15. W (step 13) = _____J/s

■ JOB SHEET 14–4 ■

Name _____

Score _____ Date _____

PERFORMANCE OBJECTIVE

Given an 80-gal storage tank, four collectors (approximately 80 ft^2), a sunny day, and a temperature recorder, determine the number of hours and the quantity of solar energy required to raise the temperature of the storage-tank water 75°F.

REFERENCE

Heating, Ventilating, and Air Conditioning Fundamentals (Chapter 14)

EQUIPMENT

Installation similar to Figure 14–4 with 80-gal storage tank; temperature recorder, electronic temperature thermometer, calculator, notebook

JOB 14–4

Calculating sun energy (helio) and number of hours of sunshine required to raise temperature of storage-tank water 75°F

PROCEDURE

1. Fill solar system with city water.
2. Place bulb of temperature recorder in storage-tank thermometer well.
3. Calculate total heat to raise storage-tank water (Btu = weight × specific heat × temperature rise).
    8.33 lb/gal × 80 gal × 1 × 75 = _____Btu
4. Add tank heat loss during heating cycle (5 percent approximate). Btu (step 3) plus 5 percent _____Btu
5. Determine helio energy (sunlight striking collector array) required for conversion to heat energy (thermal). Overall efficiency of collectors generally 50 percent.
    Btu (step 4) ÷ 0.5 = _____Btu
6. With electronic thermometer, measure temperature rise of water entering and leaving solar collector array: _____°F rise
7. Install in-line flow meter and record flow rate: _____gal/min
8. Find insolation of collectors per hour.
    500 × gal/min (step 7) × temperature rise (step 6) = _____Btu/h
9. Find insulation per 8-h sunny day.
    (8 × step 8) ÷ 80 ft^2 (collector area) = _____Btu(ft^2)(day)
10. Determine the number of sun energy hours to heat the water.
    step 5 ÷ step 8 = _____h
11. Storage-tank temperature-recording chart shows: _____°F rise.
12. The average temperature rise of the storage tank per hour (8-h day): _____°F.

■ JOB SHEET 14–5 ■

Name _____

Score _____ Date _____

PERFORMANCE OBJECTIVE

Given a 30-gal tank of tap water, insulated cabinet, black paint, and tin foil, build a passive-type water heater. If properly oriented on a clear day, a minimum of 30° rise in temperature will be obtained.

REFERENCE

Heating, Ventilating, and Air Conditioning Fundamentals (Chapter 14)

EQUIPMENT

Discarded hot-water tank with insulation removed, discarded refrigerator large enough to hold stripped water tank

TOOLS

Pipe wrench, hand tools, electronic three-lead temperature tester

SUPPLIES

Garden hose, tin foil, duct tape, black spray paint, hose bib

JOB 14–5

Making a passive solar water heater

PROCEDURE

1. Remove cover and insulation from water tank.
2. Plug the normal hot-water tank cold-water inlet connection.
3. Paint the tank black.
4. Spray the refrigerator interior black.
5. Line the inside door of the refrigerator with tin foil.
6. Remove condensing unit from the refrigerator and place on back side.
7. Place hot-water tank in refrigerator cabinet with drain valve facing door opening.
8. Connect a water inlet line with a hose bib attached to the normal hot-water outlet connection.
9. Properly orient tank outdoors in sunlight (due south).
10. Tilt refrigerator for latitude degree (refer to Figure A–3).
11. Fill tank (through normal outlet connection).
12. Cover door opening of refrigerator cabinet with a sheet of plastic or glass and record temperature rise at tank drain valve.

■ JOB SHEET 14-6 ■

Name _____

Score _____ Date _____

PERFORMANCE OBJECTIVE
Given an insulated storage tank, flat-plate collector(s), and sufficient pipe and fittings for water connections, build a thermosyphon water heater.

REFERENCE
Heating, Ventilating, and Air Conditioning Fundamentals (Chapter 14); Figures 14–7 and 14–10

EQUIPMENT
Flat-plate solar collector, storage tank

TOOLS
Pipe wrenches, hand tools, compass, protractor

SUPPLIES
PVC (polyvinylchloride), galvanized pipe, or copper tubing, two check valves, six shut-off valves, tempering valve, air vent, vacuum relief, high-pressure relief

JOB 14–6
Building a thermosyphon water heater

PROCEDURE
1. Orient collector array.
2. Install insulated storage tank 2 ft (0.6 m) above solar collector (Figure 14–8).
3. Connect the piping as shown in Figure 14–7*b*.
4. If a standard water tank is used:

   a. Collector outlet connects to cold-water inlet.
   b. Hot water is taken from normal hot-water outlet (top of tank).
   c. Remove boiler drain valve and install a tee.
   d. Connect cold-water supply to tee (step *c*).
   e. Connect collector inlet to tee (step *c*).

5. Connect temperature sensor to inlet and outlet piping of collector, and third lead to hot-water outlet pipe.

■ JOB SHEET 14–7 ■

Name _____

Score _____ Date _____

PERFORMANCE OBJECTIVE
Given the text data, drafting tools, and tape measure, lay out on
paper a solar preheat domestic water system that employs the thermosyphon principle.

REFERENCE
Heating, Ventilating, and Air Conditioning Fundamentals (Chapters 8 and 14); Unit 8, "Piping and Line Sizing"; basic drafting book

EQUIPMENT
Drafting machine (optional), job site

TOOLS
Typical drafting tools used for introductory drafting course, tape
measure

SUPPLIES
Sketch pad with graph lines, map giving latitude degrees of job site,
plan drawing of residence (optional)

JOB 14–7
Making a scale drawing of a thermosyphon domestic water pre-
heater for a residence or school shower-room water heater

PROCEDURE
1. Review the text material.
2. Select job site (school or home).
3. Determine the number of collectors and a due-south installation site.
4. Determine tilt angle for collectors. Find latitude listed in Figure A-3 and add 15°.
5. Determine how to run piping from collector storage to existing heater.
6. Draw a rough sketch of complete job (not drawn to scale).
7. Indicate lengths of pipe needed for all runs.
8. Draw an isometric drawing from sketch.
9. Number all components (see Figure 14–4) and identify all
   numbers in the legend.

■ JOB SHEET 14–8 ■

Name _____

Score _____ Date _____

PERFORMANCE OBJECTIVE
Given the text data, drafting tools, and a tape measure, make an isometric drawing of a domestic solar water heater with an automatic drain-down system, using your home as the job site. Scaled drawing will be applicable with minor errors.

REFERENCE
Heating, Ventilating, and Air Conditioning Fundamentals (Chapters 8 and 14); Figure 14–4; any basic drafting book; Unit 8, "Piping and Line Sizing"

EQUIPMENT
Drafting machine (optional), job site

TOOLS
Tools required for an introductory drafting class, tape measure

SUPPLIES
Sketch pad

JOB 14–8
Making an isometric drawing of solar-heated domestic water system with automatic drain down

PROCEDURE
1. Refer to text information. (See Figure 14–4.)
2. Determine the collector area required (20 ft^2/person).
3. Size storage tank (gal/ft^2 of collector).
4. Determine pipe size.
5. Select collector location (due south) at job site.
6. Sketch piping layout.
7. Dimension runs of pipe.
8. From sketch information draw an isometric drawing of the system.
9. Make a schematic electrical diagram.

■ JOB SHEET 14–9 ■

Name _____

Score _____ Date _____

PERFORMANCE OBJECTIVE
Given the swimming-pool area in square feet, a photograph of residence with a pool, or a field trip to job site, list the pool-collector panels required for the job. Itemize the solar hardware required for the installation. Make a shop drawing of piping and components.

REFERENCE
Heating, Ventilating, and Air Conditioning Fundamentals (Chapters 8 and 14); Figures 14–8 and 14–9

EQUIPMENT
Residence with a swimming pool (actual or illustrated drawing)

TOOLS
Drafting tools, tape measure

SUPPLIES
Sketch pad

JOB 14–9
Making a shop drawing of a solar pool-heater installation

PROCEDURE
1. Survey installation site.
2. Determine square feet of pool surface area.
3. Select panel positioning for maximum efficiency: garage roof, patio cover, or open slope.
4. Determine square feet of collector area using the following criteria:
   a. Panels facing south: panel area minimum of 50 percent pool area—70 percent pool area recommended.
   b. Panel facing east or west: minimum of 70 percent pool area—85 percent is recommended.
5. For panels facing south on a flat roof, figure 70 percent of pool area. All panels must have a minimum of 5° slope from the bottom inlet to the top outlet to ensure proper system venting.
6. Determine number of collectors:
   a. Sizing factor times square feet (steps 4 and 5) = net area.
   b. Divide net area by nominal area of single collector.
7. Refer to system diagram for automatic control (Figure 14–9).
8. Make a shop drawing of piping arrangement.
9. Dimension drawing.

10. Polyvinyl chloride (PVC) piping may be substituted for copper; however, run copper stub out from collector [minimum 10 ft (3.04 m)] because excessively high temperatures occur within a few feet of collector area that can damage PVC.

11. List material and fittings required to complete job.

■ MULTIPLE-CHOICE TEST 1 ■

Name _____

Score _____ Date _____

DIRECTIONS
Circle the letter that best answers the following multiple-choice questions.

1. The solar collector most popular for residential applications is the:

   a. flat plate   c. round plate
   b. concave plate   d. parabolic

2. Solar collectors should be oriented due:

   a. north   c. east
   b. south   d. west

3. The optimum tilt angle for solar collectors is latitude plus 15°:

   a. north   c. east
   b. south   d. west

4. Solar glazing should have a low:

   a. plastic content
   b. copper content
   c. magnesium content
   d. iron content

5. Solar collectors are constructed with:

   a. no glazing
   b. single glazing
   c. double glazing
   d. all the above

6. A solar rock storage-bin construction requires:

   a. a top plenum
   b. a bottom plenum

   c. a top and bottom plenum
   d. a solid washed-rock plenum

7. Glycol solutions are commonly used in:
   a. low ambient air-to-air systems
   b. closed hydronic systems
   c. open hydronic systems
   d. drain-down water systems

8. In a hydronic drain-down system, water returning to the collectors is taken from the:

   a. bottom of the storage tank
   b. top of the storage tank
   c. center of the storage tank
   d. top or bottom of the storage tank

9. A closed hydronic system must have either a diaphragm or air-cushion:

   a. vacuum-relief valve
   b. automatic vent
   c. pressure regulator
   d. expansion tank

10. The glycol solution pump for residential solar applications is generally constructed of stainless steel and:

   a. 1/20 hp
   b. 1/3 hp
   c. 1/2 hp
   d. 0.75 hp

■ WRITTEN EVALUATION 1 ■

1. You are given a truckload of flat-plate solar collectors to deliver to a job and install. What technical information and special tools do you need to complete the job?
2. Two residential solar installations with air collectors are installed on the same street. You are told that one works poorly compared with the other and that the problem is probably due to the rock storage bin. What can go wrong with a box full of rocks?
3. Make a list of components that are used in a closed hydronic solar system and not necessarily found in a drain-down system. Also, give a reason for the need for each of these components.

■ MULTIPLE-CHOICE TEST 2 ■

Name _____

Score _____ Date _____

DIRECTIONS
Circle the letter that best answers the following multiple-choice questions.

1. Passive solar systems:
   a. are direct-energy storage systems
   b. are indirect-energy storage systems
   c. utilize a separate rock bed or storage tank for thermal storage
   d. rely on forced convection and radiation

2. By implementing the proper design principles, seasonal performance costs of heating and cooling can be reduced up to:
   a. 20 percent   c. 40 percent
   b. 30 percent   d. 50 percent

3. Degree days are fahrenheit degrees difference between the mean temperature of the day and a base temperature of:
   a. 45°F   c. 65°F
   b. 55°F   d. 75°F

4. The U factor is:
   a. the reciprocal of the insulation resistance factor
   b. increased when the number of degree days is lowered
   c. found by dividing the R factor by 10
   d. usually greater than 1

5. Solar cooling requires high-efficiency collectors and preferably:
   a. standard glazing
   b. water-white glazing
   c. double-pane glazing
   d. triple-pane glazing

6. One of the following does not apply to passive solar systems:
   a. simplicity
   b. low cost
   c. unreliability
   d. opportunity for creativity

7. Most of the windows should be placed:
   a. north side   c. east side
   b. south side   d. west side

8. A normal roof overhang:
   a. will block out the summer sun
   b. will block out the winter sun
   c. lowers the efficiency of a Kalwall heater
   d. promotes nocturnal cooling

9. The roof of a house should be covered with:
   a. a light-color surface material
   b. a dark-color surface material
   c. heavy-duty shingles
   d. heat-conducting material

10. A solar wall heater can be turned off by:
    a. a reflector
    c. a thermostat
    b. an outside vent
    d. both *a* and *b*

■ WRITTEN EVALUATION 2 ■

1. What would be required to install a sun wall in your home? If you have a sun wall, what can be done to increase its performance?

2. Review the passive solar design principles listed in the text and make a report on how your home or apartment building matches up to the five principles.

3. If the three-way mixing valve were left out of the thermosyphon system, what would probably happen, and what changes would be required in the cold-water piping?

■ MULTIPLE-CHOICE TEST 3 ■

DIRECTIONS
Circle the letter that best answers the following
multiple-choice questions.

1. An active solar system can be described as:

   a. a direct system
   b. an indirect system
   c. an overpassive system
   d. a natural-convection system

2. Manufacturers estimate a 20-year life expectancy on their solar hardware but:

   a. solar equipment will be winterized and
      need replacement before 20 years
   b. the life expectancy should exceed 20
      years
   c. the equipment should be amortized in
      approximately 10 years
   d. the equipment cannot be amortized
      until the last mortgage payment is
      made

3. The solar index:

   a. rates equipment on a scale from 1 to 10
   b. and the pollen count coincide
   c. indicates the percentage of hot water
      that could be supplied on a given day
   d. was originated by President Carter for
      the purpose of giving homeowners tax
      credits ranging from $400 to $2,200

4. A properly piped drain-down system:

   a. requires a 25 percent glycol solution
   b. requires a 50 percent glycol solution
   c. requires a glycol solution 20° below
      the degree day base of the installation
      site
   d. does not require a glycol solution

5. An installation with 20 flat-plate water
   collectors or 400 ft^2 (36.8 m^2) should have
   a storage tank with a capacity of:

   a. 60 gal (227 L)   c. 100 gal (378.5 L)
   b. 80 gal (302 L)   d. 400 gal (1,514 L)

6. Opposed damper blades travel 90° from
   open to closed position while the damper
   motor shaft travels:

   a. 90°
   b. 160 or 180°
   c. 360°
   d. 180 r/min

7. An air gap on a storage-tank fill line:

   a. is used to prevent solar water from sy-
      phoning back into the makeup line
   b. prevents the tank from becoming
      water-logged
   c. relieves unwanted oxygen from the
      system
   d. is needed to prime the pump

8. The following is a solar-powered unit:

   a. augmented heat pump
   b. Suncell turbine
   c. Arkla lithium bromide unit
   d. Native sun compression unit

9. The following solution takes the place of
   the compressor crankcase in an absorption unit:

   a. water
   b. lithium bromide
   c. ammonia
   d. dichlorodifluoromethane

10. Normal operating pressures for a solar-
    absorption unit:

    a. 100 high, 50 low
    b. 200 high, 75 low
    c. 29.92 in Hg or 14.7 psia
    d. 47 mmHg and 7 mmHg

■ WRITTEN EVALUATION 3 ■

1. The approximate installation price of a solar-absorption system is $25,000. What factor increases the cost over a solar-augmented heat-pump system, and for what reasons will future installations probably cost less?

2. State reasons why pool collectors may cost less than collectors for space heating, and why they cannot be used for solar-cooling applications.

3. The solar-heating system described in the text that employed electrically operated dampers had four modes of operation, whereas the solar-augmented heat pump had six modes. How does the six-mode heat-pump system differ from the four-mode system described?

Unit 15

CHILLED-WATER SYSTEMS

■ INTRODUCTION

The following tasks are related to direct-expansion, flooded, absorption, and thermal-storage chilled-water systems. Quite obviously, school lab facilities could not accommodate one of each type of chiller, and more than likely, not even one because of the size of these units. Therefore, the lab tasks are designed for field trips. Each procedural step will provide an answer and a blank space to fill in the actual operating condition you observe. Any discrepancy can be accounted for by the tour guide. Be sure to read the text material (Chapter 15) prior to the field trip.

■ JOB SHEET 15–1 ■

Name _____

Score _____ Date _____

PERFORMANCE OBJECTIVE
Given an operating chiller with a reciprocating compressor and hydraulic unloaders, fill in the blanks for each procedural step and compare your findings with given answers. The person leading the tour should be able to explain any discrepancies you can't account for.

REFERENCE
Heating, Ventilating, and Air Conditioning Fundamentals (Chapter 15)

TOOLS
Hand tools, thermometers, VOM, amprobe, gauge manifold, pressure temperature chart

MATERIAL
Notebook and pencil

JOB 15–1
Direct expansion chiller–AZ 20 refrigerant performance analysis, air-conditioning application

PROCEDURE
1. Condensing temperature parameters: 86°F to 130°F (30°C to 54°C)
2. Refrigerant in system: AZ 20 or_____
3. Head pressure, water-cooled condenser: 132 psig
   Gauge reading: _____psig
4. Suction pressure: 32 psig to 39 psig (32 to 40°F).
   Gauge reading: _____psig.
5. Oil pressure: minimum 10 psig; maximum 50 psig
   Suction pressure minus oil-pressure gauge = _____psig
6. Cylinder unloaders, fully unloaded at 32°F suction: 32 psig
   Throttle suction service valve to check unloaders
   Unloaded: _____psig _____amps compressor motor
   Loaded: _____psig _____amps
7. Discharge line (6 in from compressor) maximum: 220°F (104°C)
   Temperature recorded: _____°F
8. Suction line (6 in from compressor) maximum: 65°F (18°C)
   Temperature recorded: _____°F
9. Evaporator superheat: 10 to 12°F.
   Recorded: _____°F.
10. Suction gauge pressure converted to temperature minus suction temperature at compressor = 15 to 25 degrees
    Superheat measured: _____degrees

11. Pressure at king valve converted to temperature minus liquid line temperature = subcooling (minimum 10 degrees)
Record subcooling: _____degrees
12. Temperature drop across drier = 1 degree (2 psig)
Feel with your hands the drier inlet and outlet (no difference).
Measured difference: _____degrees F.
13. Motor horsepower rating: _____
14. Full load amps: _____
15. Running load amps: _____
16. Brake horsepower (RLA / FLA × HP): _____BHP

■ JOB SHEET 15–2 ■

Name _____

Score _____ Date _____

PERFORMANCE OBJECTIVE
Convert an R-11 flooded chiller to refrigerant R-123 and make a performance comparison. Operating performance will match the refrigerant manufacturer's thermodynamic tables.

REFERENCE
Heating, Ventilating, and Air Conditioning Fundamentals (Chapter 15); Allied Signal, technical bulletin #646

TOOLS
Recovery unit, hand tools, oil pump, brazing torch, tube cutter, file, VOM

MATERIAL
Refrigerant and oil

JOB 15–2
Convert R-11 chiller to R-123

PROCEDURE
1. Record operating conditions and compare to operating engineer's log. Are there any changes?_____
2. Observe the coolers' refrigerant sight glass. Does the refrigerant cover the tube bundle?_____
3. Discharge pressure: _____psig; Temperature: _____°F; _____°C
4. Suction pressure: _____psig; Temperature _____°F; _____°C
5. Compression ratio: _____ _____.
6. Oil pressure: _____psig
7. Compressor oil level: _____High _____Low _____OK
8. Oil temperature (25 above condensing): _____°F
9. Test oil heaters: _____; Oil cooler: _____
10. Check purge recovery unit (relief valve must be vented outdoors). If unit does not meet code regulations it must be replaced.
11. Condenser water:
    Inlet pressure: _____psig; Temperature: _____°F
    Outlet pressure: _____psig; Temperature: _____°F
12. Drain oil from purge unit compressor and oil separator. Recharge with the proper oil (Tables 15–1 and 15–2). Type oil: _____
13. Run purge unit to remove noncondensibles.
14. Select proper oil (Tables 15–1 and 15–2): _____type
15. Change compressor oil: _____gallons
16. Run unit 48 hours and change oil.

Typical Properties of EMKARATE™ RL Polyolester Refrigeration Lubricants

OEM recommendations should be followed, when available, to ensure that the correct oil is used and that the compressor or system warranty is not invalidated.

Catalog Number*	Container Size	Viscosity @104 F (40 C)	@212 F (100 C)	@-4 F (-20 C)	Viscosity Index	Pour Point	Miscibility High	Miscibility Low	Herm. Recip.	Herm. Rotary	Auto. Recip.	Auto. Rotary Vane	Auto. Scroll	Ind. Recip.	Ind. Centrifugal	Ind. Screw	Ind. Scroll
LE22H1 / LE22H5 / LE22H55	1 Gallon / 5 Gallon / 55 Gallon	19.9	4.3	505	128	-62 F -52 C	>176 F (80 C)	-47 F (-44 C)	•					•	•		
LE32SQ / LE32S1 / LE32S5 / LE32S55	1 Quart / 1 Gallon / 5 Gallon / 55 Gallon	32.0	5.6	2310	114	-51 F -46 C	>176 F (80 C)	+23 F (-5 C)	•	•			•	•	•		
LE68H1 / LE68H5 / LE68H55	1 Gallon / 5 Gallon / 55 Gallon	74.1	10.1	7356	118	-31 F -35 C	>176 F (80 C)	-29 F (-34 C)		•	•	•	•	•	•	•	•
LE32H1 / LE32H5 / LE32H55	1 Gallon / 5 Gallon / 55 Gallon	32.0	5.6	2310	114	-62 F -52 C	>176 F (80 C)	-29 F (-34 C)	•					•	•		
LE32SB1 / LE32SB5 / LE32SB55	1 Gallon / 5 Gallon / 55 Gallon	32.0	5.6	2310	114	-51 F -46 C	>176 F (80 C)	+23 F (-5 C)	•	•			•	•	•		
LE46H1 / LE46H5 / LE46H55	1 Gallon / 5 Gallon / 55 Gallon	48.0	7.3	3649	111	-40 F -40 C	>176 F (80 C)	-29 F (-34 C)		•		•	•	•	•	•	•

(Primary Viscosities: first three rows. Specialty Viscosities (OEM Specific): last three rows.)

* Catalog number indicates viscosity.

TABLE 15–1   Polyolester lubricants.

Physical Property	Test Method	AB150	AB200A	AB300
Viscosity, cSt at 40 C (104 F)	D445	32	46	68
Pour Point, Maximum	D97	-49 F (-45 C)	-31 F (-35 C)	-31 F (-35 C)
Flash Point, Minimum	D92	302 F (150 C)	329 F (165 C)	329 F (165 C)
Floc Point	ASHRAE 86	-85 F (-65 C)	-67 F (-55 C)	-67 F (-55 C)
Dielectric Strength, kV, Min.	D877	30	30	30
Specific Gravity, Minimum	D1298	0.86	0.86	0.86
Color, Maximum	D1500	1	1	1
Acid No., Mg KOH/gm, max.	D974	0.02	0.02	0.02
Water, ppm, Maximum	D1533	30	30	30

Grade	SUS Viscosity	ISO Viscosity	1 Gallon	5 Gallon	55 Gallon
AB150	150	32	LAB15-1	LAB15-5	LAB15-55
AB200A	200	46	LAB20-1	LAB20-5	LAB20-55
AB300	300	68	LAB30-1	LAB30-5	LAB30-55

TABLE 15–2   Polyalkylene glycol (PAG) and alkylated benzene lubricants.

17. Repeat step 16 and test oil for mineral oil content; mineral oil content should be 1 percent or less.
18. Remove R-11 and charge the system with R-123.
19. Refer to Table 15–3 and draw a cycle diagram.
20. Compare cycle performance with Figure 15–3 from the textbook and actual operating conditions (performance should match).

TABLE 15–3 Thermodynamic table R123.

Genetron® 123
(Dichlorotrifluoroethane)

GENETRON® 123 is a very low-ozone-depleting compound that serves as a replacement to CFC-11 in centrifugal chillers.

Physical Properties:

Chemical Formula	$CHCl_2CF_3$
Molecular Weight	152.9
Boiling Point @ 1 Atm (°F)	82.2
Critical Temperature (°F)	363
Critical Pressure (Psia)	540
Critical Density (lb./cu. ft.)..................	34.5
Saturated Liquid Density @ 86°F (lb./cu. ft.)......	90.4
Heat of Vaporization at Boiling Point (Btu/lb)	72.9
Specific Heat of Liquid @ 86°F (Btu/lb °F)	0.21
Specific Heat of Vapor @ Constant Pressure	0.17
(Cp @ 86°F and 1 Atm, Btu/lb °F)	
Flammable range, % volume in air	Nonflammable
(based on ASHRAE Standard 34 with match ignition)	
Ozone Depletion Potential..................	0.016

Comparative Cycle Performance

Evaporator temperature: 35°F
Condenser temperature: 105°F
Degrees superheat @ evaporator: 0°F
Degrees subcooling: 0°F
Compressor isentropic efficiency: 75%

	Genetron®	
	123	11
Evaporator pressure, in Hg	19.5	17.2
Condenser pressure, psig	8.1	10.9
Compression ratio	4.47	4.06
Compressor discharge temperature, °F	122.8	144.0
Coefficient of performance	4.63	4.72
Refrigerant circulation per ton, lb./min.	3.29	3.01
Compressor displacement per ton, cfm	21.78	18.20
Liquid flow per ton, cu. in./min.	64.1	57.9
Latent heat at evaporator temp., Btu/lb.	76.9	81.0
Net refrigeration effect, Btu/lb.	60.7	66.4

123 Thermodynamic Table

Temp. (°F)	Pressure (Psia)	Density (lb/ft³)	Vapor Volume (ft³/1b)	H_{liq} (Btu/lb)	Enthalpy ΔH_{vap} (Btu/lb)
0	2.00	97.71	15.9382	8.16	79.40
2	2.12	97.54	15.0945	8.58	79.26
4	2.25	97.38	14.3033	9.00	79.12
6	2.38	97.22	13.5607	9.43	78.97
8	2.52	97.05	12.8634	9.85	78.84
10	2.66	96.89	12.2082	10.27	78.70
12	2.82	96.72	11.5923	10.70	78.55
14	2.97	96.56	11.0131	11.13	78.41
16	3.14	96.39	10.4679	11.56	78.26
18	3.31	96.23	9.9545	11.99	78.11
20	3.50	96.06	9.4709	12.42	77.97
22	3.69	95.90	9.0150	12.85	77.82
24	3.88	95.73	8.5851	13.28	77.68
26	4.09	95.56	8.1794	13.72	77.52
28	4.31	95.40	7.7964	14.15	77.38
30	4.53	95.23	7.4347	14.59	77.22
32	4.77	95.06	7.0929	15.03	77.07
34	5.01	94.90	6.7697	15.46	76.92
36	5.26	94.73	6.4639	15.90	76.77
38	5.53	94.56	6.1746	16.35	76.61
40	5.80	94.39	5.9007	16.79	76.45
42	6.09	94.22	5.6412	17.23	76.30
44	6.39	94.05	5.3953	17.68	76.13
46	6.70	93.88	5.1622	18.12	75.98
48	7.02	93.71	4.9411	18.57	75.82
50	7.35	93.54	4.7312	19.02	75.65
52	7.70	93.37	4.5320	19.47	75.49
54	8.06	93.20	4.3429	19.92	75.33
56	8.43	93.03	4.1631	20.37	75.16
58	8.82	92.86	3.9922	20.82	75.00
60	9.22	92.69	3.8297	21.28	74.83
62	9.63	92.51	3.6752	21.73	74.66
64	10.06	92.34	3.5281	22.19	74.49
66	10.51	92.17	3.3880	22.64	74.33
68	10.97	91.99	3.2546	23.10	74.16
70	11.44	91.82	3.1275	23.56	73.98
72	11.94	91.65	3.0064	24.02	73.81
74	12.44	91.47	2.8908	24.49	73.63
76	12.97	91.30	2.7806	24.95	73.45
78	13.52	91.12	2.6755	25.41	73.28
80	14.08	90.94	2.5751	25.88	73.10
82	14.66	90.77	2.4793	26.35	72.91
84	15.26	90.59	2.3877	26.81	72.74
86	15.87	90.41	2.3002	27.28	72.55
88	16.51	90.24	2.2165	27.75	72.37
90	17.17	90.06	2.1366	28.22	72.19
92	17.85	89.88	2.0600	28.70	71.99
94	18.55	89.70	1.9868	29.17	71.81
96	19.27	89.52	1.9167	29.64	71.62
98	20.01	89.34	1.8496	30.12	71.43
100	20.77	89.16	1.7853	30.59	71.24
102	21.56	88.98	1.7237	31.07	71.05
104	22.37	88.79	1.6647	31.55	70.85
106	23.20	88.61	1.6081	32.03	70.66
108	24.06	88.43	1.5538	32.51	70.46
110	24.94	88.25	1.5017	32.99	70.27
112	25.85	88.06	1.4518	33.48	70.06
114	26.78	87.88	1.4038	33.96	69.86
116	27.74	87.69	1.3577	34.45	69.66
118	28.72	87.51	1.3135	34.93	69.46
120	29.74	87.32	1.2710	35.42	69.25

TABLE 15–3 (Continued)

123 Thermodynamic Table (continued)

Enthalpy H_{vap} (Btu/lb)	Entropy S_{liq} (Btu/lb. °F)	Entropy S_{vap} (Btu/lb °F)
87.56	0.0186	0.1913
87.84	0.0195	0.1911
88.12	0.0204	0.1910
88.40	0.0213	0.1909
88.69	0.0222	0.1908
88.97	0.0231	0.1907
89.25	0.0240	0.1905
89.54	0.0249	0.1904
89.82	0.0258	0.1903
90.10	0.0267	0.1903
90.39	0.0276	0.1902
90.67	0.0285	0.1901
90.96	0.0294	0.1900
91.24	0.0303	0.1899
91.53	0.0312	0.1899
91.81	0.0321	0.1898
92.10	0.0330	0.1897
92.38	0.0339	0.1897
92.67	0.0348	0.1896
92.96	0.0356	0.1896
93.24	0.0365	0.1895
93.53	0.0374	0.1895
93.81	0.0383	0.1895
94.10	0.0392	0.1894
94.39	0.0401	0.1894
94.67	0.0409	0.1894
94.96	0.0418	0.1894
95.25	0.0427	0.1894
95.53	0.0436	0.1893
95.82	0.0445	0.1893
96.11	0.0453	0.1893
96.39	0.0462	0.1893
96.68	0.0471	0.1893
96.97	0.0479	0.1893
97.26	0.0488	0.1893
97.54	0.0497	0.1893
97.83	0.0505	0.1894
98.12	0.0514	0.1894
98.40	0.0523	0.1894
98.69	0.0531	0.1894
98.98	0.0540	0.1894
99.26	0.0549	0.1895
99.55	0.0557	0.1895
99.83	0.0566	0.1895
100.12	0.0574	0.1896
100.41	0.0583	0.1896
100.69	0.0591	0.1897
100.98	0.0600	0.1897
101.26	0.0609	0.1897
101.55	0.0617	0.1898
101.83	0.0626	0.1898
102.12	0.0634	0.1899
102.40	0.0643	0.1899
102.69	0.0651	0.1900
102.97	0.0659	0.1901
103.26	0.0668	0.1901
103.54	0.0676	0.1902
103.82	0.0685	0.1903
104.11	0.0693	0.1903
104.39	0.0702	0.1904
104.67	0.0710	0.1905

Temp. (°F)	Pressure (Psia)	Liquid Density (lb/ft³)	Vapor Volume (ft³/1b)	Enthalpy H_{liq} (Btu/lb)	Enthalpy ΔH_{vap} (Btu/lb)	Enthalpy H_{vap} (Btu/lb)	Entropy S_{liq} (Btu/lb °F)	Entropy S_{vap} (Btu/lb °F)
122	30.78	87.13	1.2301	35.91	69.05	104.96	0.0718	0.1905
124	31.84	86.94	1.1909	36.40	68.84	105.24	0.0727	0.1906
126	32.94	86.76	1.1531	36.89	68.63	105.52	0.0735	0.1907
128	34.06	86.57	1.1168	37.38	68.42	105.80	0.0743	0.1908
130	35.21	86.38	1.0818	37.88	68.20	106.08	0.0752	0.1908
132	36.40	86.19	1.0482	38.37	67.99	106.36	0.0760	0.1909
134	37.61	86.00	1.0158	38.87	67.78	106.65	0.0768	0.1910
136	38.85	85.81	0.9847	39.36	67.57	106.93	0.0777	0.1911
138	40.13	85.61	0.9547	39.86	67.35	107.21	0.0785	0.1912
140	41.44	85.42	0.9257	40.36	67.12	107.48	0.0793	0.1913
142	42.78	85.23	0.8979	40.86	66.90	107.76	0.0802	0.1914
144	44.15	85.03	0.8710	41.36	66.68	108.04	0.0810	0.1914
146	45.56	84.84	0.8451	41.86	66.46	108.32	0.0818	0.1915
148	47.00	84.64	0.8201	42.37	66.23	108.60	0.0826	0.1916
150	48.47	84.44	0.7960	42.87	66.01	108.88	0.0835	0.1917
152	49.98	84.25	0.7728	43.38	65.77	109.15	0.0843	0.1918
154	51.53	84.05	0.7504	43.88	65.55	109.43	0.0851	0.1919
156	53.11	83.85	0.7287	44.39	65.32	109.71	0.0859	0.1920
158	54.73	83.65	0.7078	44.90	65.08	109.98	0.0867	0.1921
160	56.38	83.45	0.6876	45.41	64.85	110.26	0.0876	0.1922

123 Thermodynamic Formulas

T_c =363.200 °F P_c =533.097 psia ρ_c =34.5257 lb./cu.ft. T_b =82.166 °F $MWt.$ =152.930

Experimental vapor pressure correlated as:

$$\ln(P_{vap}) = A + \frac{B}{T} + CT + DT^2 + \frac{E(F-T)}{T}\ln(F-T)$$

where P_{vap} is in psia and T in °R

A=0.2135167313E+02 B=−0.7580945477E+04 C=−0.1151736692E−01
D=0.5341983248E−05 E=0.0000000000E+00 F=0.0000000000E+00

Experimental ideal gas heat capacity correlated as:

$$C_p^o \text{ (Btu/lb. °R)} = C_1 + C_2 T + C_3 T^2 + C_4 T^3 + C_5/T$$

where T is in °R

C_1=0.3627324125E−01 C_2=0.2963321983E−03 C_3=−0.1222965602E−06
C_4=0.0000000000E+00 C_5=0.0000000000E+00

Experimental liquid density correlated as:

$$\rho = \rho_c + \sum_{i=1}^{4} D_i (1 - T_r)^{i/3}$$

where ρ is in lb./cu.ft.

D_1=0.5473153636E+02 D_2=0.6881690823E+02 D_3=−0.9265622670E+02 D_4=0.6699838557E+02
ρ_c =0.3452572608E+02

Estimated Martin-Hou coefficients used:

$$P = \frac{RT}{(v-b)} + \sum_{i=2}^{5} \frac{A_i + B_i T + C_i\, e^{(-KT_r)}}{(v-b)^i}$$

P (psia), v (cu.ft./lb.), T (°R), $T_r = T/T_c$

R =0.070173 b =0.5778313758E−02 K=0.5474999905+01

i	A_i	B_i	C_i
2	−0.3461174842E+01	0.1482683303E−02	−0.6375783935E+02
3	0.1271057059E+00	−0.9675560464E−04	0.1712913479E+01
4	−0.5983292209E−03	0.0000000000E+00	0.0000000000E+00
5	−0.7744198272E−05	0.1325431541E−07	−0.1261428005E−03

■ JOB SHEET 15–3 ■

Name _____

Score _____ Date _____

PERFORMANCE OBJECTIVES
Given an operating absorption chiller and the operating engineer's log book, confirm the readings and identify the main system components.

REFERENCE
Heating, Ventilating, and Air Conditioning Fundamentals (Chapter 15)

EQUIPMENT
Operating absorption chiller

TOOLS
Temperature thermocouple, pencil, and worksheet

JOB 15–3
Identify the components of an operating absorption unit and record readings.

PROCEDURE

COOLING CYCLE

1. Locate absorber.
   Chilled-water inlet temperature: _____°F; _____°C
   Chilled-water outlet temperature: _____°F; _____°C
   Cooling-water inlet temperature: _____°F; _____°C
2. Locate condenser.
   Cooling-water inlet temperature: _____°F; _____°C
   Cooling-water outlet temperature: _____°F; _____°C
3. Locate and activate purge unit if needed.
   Record evaporator temperature: _____ mm Hg (7mm Hg)
   Record condensing temperature: _____ mm Hg (47mm Hg)
3. Low temperature generator:
   Steam inlet temperature: _____°F; _____°C
   Steam outlet temperature: _____°F; _____°C
4. High temperature generator:
   Steam inlet temperature: _____°F; _____°C
   Steam outlet temperature: _____°F; _____°C

HEATING CYCLE

1. Hot-water inlet temperature: _____°F; _____°C
   Hot-water outlet temperature: _____°F; _____°C
2. Diluted solution temperature: _____°F; _____°C
3. Intermediate solution temperature: _____°F; _____°C

■ JOB SHEET 15–4 ■

Name _____

Score _____ Date _____

PERFORMANCE OBJECTIVE
Plot an ammonia (R-717) cycle of a basic ice storage system on a
pressure/enthalpy skeletal diagram. Compare the theoretical cycle
to conditions of an operating unit.

REFERENCE
Heating, Ventilating, and Air Conditioning Fundamentals (Chap-
ter 15); Baltimore Aircoil, *A Guide to Ice Storage System Design;*
ASHRAE thermodynamic tables

MATERIAL
P/E chart, thermodynamic tables, ruler, pencil, and paper

JOB 15–4
Plot a basic ice storage system on a pressure/enthalpy diagram. Ex-
cerpt data from Figure 15–1 and check your answers with Table
15–4

PROCEDURE
1. Subcooled liquid leaving evaporative condenser at 85°F at
   point *A*. Operating unit: _____°F; _____°C
2. Constant quality, Point *B*: _____% (vapor)
3. Flash gas, *B-C*: _____btu/lb
4. Net refrigeration effect, *B-D*: _____btu/lb
5. Point *D*, constant volume: _____cubic feet/lb
6. Point *E* evaporator outlet, 5 degree superheat: _____
7. Suction line, *E-F* designed for _____degree loss.
8. Entropy, *F-G*: _____btu/lb
9. Temperature of discharge at point *G*: _____°F; _____°C
10. Desuperheat, *G-H*: _____btu/lb
11. Latent heat of condensation, *H-I*: _____btu/lb
12. Subcooled liquid (10 degree) *I-A*: _____btu/lb

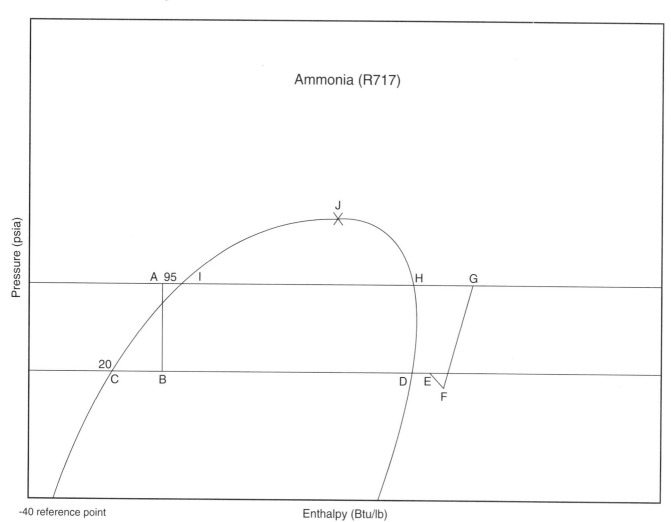

FIGURE 15–1a   Pressure-enthalpy skeletal diagram.

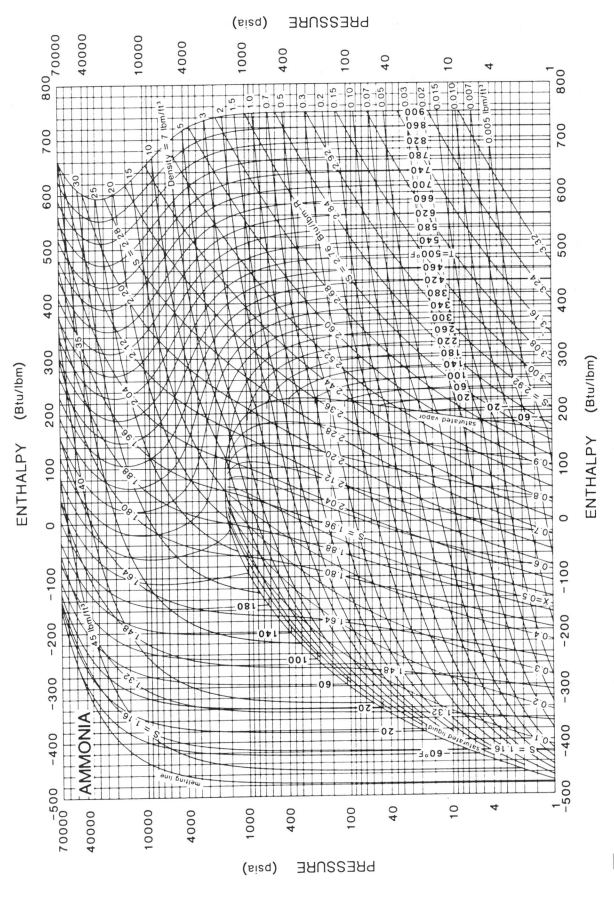

FIGURE 15–1b Pressure-enthalpy diagram for refrigerant 717 (ammonia).

239

TABLE 15–4 Refrigerant tables and charts.

Temp °F	Pressure psia	Pressure psig	Volume ft³/lb$_m$ Vapor	Density lb$_m$/ft³ Liquid	Enthalpy Btu/lb$_m$ Liquid	Enthalpy Btu/lb$_m$ Vapor	Entropy Btu/lb$_m$·°R Liquid	Entropy Btu/lb$_m$·°R Vapor
-107.81	0.87888	28.132*	251.30	45.813	-477.59	163.52	1.0046	2.8270
-100	1.2334	27.410*	182.84	45.513	-468.77	167.00	1.0294	2.7974
-90	1.8589	26.136*	124.47	45.124	-457.97	171.38	1.0590	2.7618
-80	2.7335	24.356*	86.771	44.728	-447.46	175.68	1.0871	2.7286
-70	3.9295	21.921*	61.796	44.322	-437.05	179.87	1.1141	2.6976
-60	5.5331	18.656*	44.879	43.909	-426.66	183.95	1.1404	2.6685
-55	6.5182	16.650*	38.507	43.700	-421.45	185.95	1.1534	2.6546
-50	7.6438	14.358*	33.180	43.490	-416.23	187.92	1.1662	2.6412
-45	8.9245	11.751*	28.707	43.279	-411.00	189.85	1.1789	2.6281
-40	10.376	8.7956*	24.935	43.067	-405.75	191.75	1.1914	2.6155
-35	12.015	5.4589*	21.740	42.853	-400.48	193.61	1.2039	2.6031
-30	13.859	1.7050*	19.023	42.639	-395.20	195.44	1.2162	2.5912
-28	14.658	0.0784*	18.051	42.553	-393.08	196.16	1.2212	2.5865
-27.91	14.696	0.0	18.007	42.549	-392.98	196.20	1.2214	2.5862
-26	15.493	0.7974	17.138	42.467	-390.96	196.88	1.2260	2.5818
-24	16.367	1.6715	16.280	42.381	-388.83	197.59	1.2309	2.5772
-22	17.281	2.5850	15.473	42.294	-386.71	198.29	1.2358	2.5727
-20	18.235	3.5392	14.714	42.207	-384.58	198.99	1.2406	2.5682
-18	19.232	4.5355	13.999	42.121	-382.44	199.68	1.2455	2.5638
-16	20.271	5.5752	13.325	42.034	-380.31	200.36	1.2503	2.5594
-14	21.356	6.6596	12.690	41.946	-378.17	201.04	1.2551	2.5550
-12	22.486	7.7902	12.092	41.859	-376.03	201.71	1.2599	2.5507
-10	23.664	8.9683	11.526	41.771	-373.89	202.38	1.2646	2.5464
-8	24.891	10.195	10.993	41.683	-371.74	203.04	1.2694	2.5422
-6	26.169	11.473	10.488	41.595	-369.59	203.69	1.2741	2.5380
-4	27.498	12.802	10.012	41.507	-367.43	204.33	1.2788	2.5339
-2	28.881	14.185	9.5614	41.418	-365.28	204.97	1.2835	2.5298
0	30.318	15.622	9.1351	41.329	-363.12	205.61	1.2882	2.5258
2	31.812	17.116	8.7315	41.240	-360.96	206.23	1.2929	2.5218
4	33.364	18.668	8.3493	41.151	-358.79	206.85	1.2976	2.5178
6	34.976	20.280	7.9870	41.061	-356.62	207.46	1.3022	2.5139
8	36.649	21.953	7.6436	40.972	-354.45	208.07	1.3069	2.5100
10	38.384	23.688	7.3179	40.881	-352.28	208.66	1.3115	2.5061
12	40.184	25.488	7.0088	40.791	-350.10	209.25	1.3161	2.5023
14	42.050	27.354	6.7155	40.700	-347.92	209.84	1.3207	2.4985
16	43.984	29.288	6.4367	40.609	-345.74	210.41	1.3253	2.4947
18	45.987	31.291	6.1720	40.518	-343.55	210.98	1.3298	2.4910
20	48.062	33.366	5.9202	40.426	-341.36	211.54	1.3344	2.4874
22	50.209	35.513	5.6808	40.334	-339.17	212.10	1.3389	2.4837
24	52.431	37.735	5.4530	40.241	-336.97	212.64	1.3435	2.4801
26	54.730	40.034	5.2362	40.149	-334.77	213.18	1.3480	2.4765
28	57.106	42.410	5.0297	40.056	-332.57	213.71	1.3525	2.4729
30	59.563	44.867	4.8330	39.962	-330.36	214.23	1.3570	2.4694
32	62.102	47.406	4.6455	39.868	-328.15	214.75	1.3615	2.4659
34	64.725	50.029	4.4667	39.774	-325.94	215.25	1.3659	2.4625
36	67.433	52.737	4.2963	39.679	-323.72	215.75	1.3704	2.4590
38	70.229	55.533	4.1336	39.584	-321.50	216.24	1.3748	2.4556
40	73.114	58.418	3.9783	39.489	-319.27	216.72	1.3793	2.4522
45	80.731	66.035	3.6199	39.249	-313.70	217.89	1.3903	2.4439
50	88.950	74.254	3.2997	39.006	-308.10	219.00	1.4013	2.4357
55	97.802	83.106	3.0131	38.760	-302.47	220.05	1.4122	2.4277
60	107.32	92.622	2.7560	38.511	-296.83	221.05	1.4230	2.4198
65	117.53	102.84	2.5249	38.260	-291.15	221.98	1.4338	2.4120
70	128.48	113.78	2.3166	38.005	-285.46	222.85	1.4445	2.4044
75	140.18	125.49	2.1285	37.747	-279.73	223.65	1.4551	2.3969
80	152.69	137.99	1.9585	37.486	-273.98	224.39	1.4657	2.3894
85	166.03	151.33	1.8044	37.221	-268.20	225.06	1.4763	2.3821
90	180.23	165.54	1.6644	36.952	-262.39	225.65	1.4867	2.3749
100	211.40	196.70	1.4212	36.404	-250.67	226.60	1.5076	2.3606
110	246.47	231.77	1.2187	35.840	-238.83	227.21	1.5282	2.3466
120	285.76	271.06	1.0490	35.258	-226.84	227.47	1.5488	2.3327
130	329.57	314.88	0.90588	34.657	-214.68	227.33	1.5691	2.3190
140	378.23	363.54	0.78450	34.034	-202.35	226.76	1.5894	2.3052
150	432.08	417.39	0.68093	33.388	-189.80	225.71	1.6097	2.2915
160	491.46	476.76	0.59212	32.713	-177.01	224.14	1.6300	2.2775
170	556.73	542.04	0.51553	32.007	-163.93	221.98	1.6503	2.2634
180	628.28	613.58	0.44914	31.262	-150.50	219.16	1.6708	2.2489
200	791.85	777.16	0.34048	29.628	-122.34	211.15	1.7126	2.2183
220	985.81	971.11	0.25574	27.712	-91.613	198.86	1.7566	2.1842
240	1214.6	1199.9	0.18704	25.301	-56.458	179.61	1.8053	2.1428
260	1484.5	1469.9	0.12620	21.678	-10.567	144.79	1.8669	2.0828
**270.1	1640.	1625.	0.06816	14.67	61.41	61.41	1.964	1.964

**Critical Point
*In. Hg. Vacuum
#Triple Point

Temp °F	Viscosity lb$_m$/ft·h Sat. Liquid	Viscosity lb$_m$/ft·h Sat. Vapor	Viscosity lb$_m$/ft·h Gas at p=1 atm ×10⁻²‡	Thermal Conductivity Btu/h·ft·°F Sat. Liquid	Thermal Conductivity Btu/h·ft·°F Sat. Vapor	Thermal Conductivity Btu/h·ft·°F Gas at p=1 atm ×10⁻³‡	Specific Heat c$_p$ Btu/lb$_m$·°F Sat. Liquid	Specific Heat c$_p$ Btu/lb$_m$·°F Sat. Vapor	Specific Heat c$_p$ Btu/lb$_m$·°F Gas at p=0 atm	Specific Heat c$_p$ Btu/lb$_m$·°F Gas at p=1 atm	Temp °F
-100				0.410			1.030				-100
-80				0.395			1.044				-80
-60				0.380			1.056	0.523	0.477		-60
-40				0.365			1.066	0.543	0.480		-40
-20	0.629	0.0227	2.02	0.350	0.011	11.1	1.075	0.565	0.483	0.547	-20
0	0.558	0.0237	2.11	0.335	0.012	11.7	1.083	0.590	0.486	0.536	0
20	0.494	0.0246	2.20	0.321	0.012	12.2	1.092	0.620	0.489	0.528	20
40	0.437	0.0256	2.29	0.306	0.013	12.9	1.103	0.655	0.493	0.522	40
60	0.386	0.0266	2.39	0.291	0.015	13.5	1.118	0.698	0.496	0.519	60
80	0.341	0.0277	2.48	0.276	0.016	14.2	1.135	0.750	0.500	0.517	80
100	0.301	0.0288	2.58	0.261	0.018	14.9	1.158	0.814	0.504	0.517	100
120	0.268	0.0299	2.67	0.246	0.020	15.6	1.187	0.866	0.508	0.518	120
140	0.238	0.0312	2.77	0.231	0.022	16.3	1.222		0.512	0.520	140
160	0.213	0.0325	2.87	0.216	0.025	17.1	1.265		0.517	0.523	160
180	0.190	0.0340	2.97	0.201	0.029	17.8	1.317		0.521	0.527	180
200	0.171	0.0358	3.06	0.186	0.034	18.6	1.379		0.526	0.532	200
220	0.154	0.0380	3.16	0.169	0.039	19.4	1.452		0.530	0.536	220
240	0.124	0.0411	3.26	0.149	0.044	20.2	1.536		0.535	0.542	240
250	0.110	0.043	3.31	0.137	0.049	20.7	1.59		0.537	0.544	250
260	0.094	0.046	3.36	0.120	0.060	21.1	1.65		0.540	0.547	260
270			3.41	0.102	0.081	21.5			0.542	0.550	270
271*	0.060	0.060	3.42	0.087	0.087	21.5			0.542	0.550	271*
280			3.46			21.9			0.544	0.552	280
290			3.51			22.3			0.547	0.555	290
300			3.56			22.8			0.549	0.557	300
320			3.66			23.7			0.554	0.562	320
360			3.86			25.4			0.564	0.572	360
400			4.06			27.3			0.574	0.581	400
440			4.27			29.1			0.584	0.590	440
500			4.57			32.0			0.599	0.602	500

*Critical Temperature. Tabulated properties ignore critical region effects.   ‡Actual value = (Table value) × (Indicated multiplier).

■ MULTIPLE-CHOICE TEST ■

Name _____

Score _____ Date _____

DIRECTION

Circle the letter that best answers the following multiple-choice questions.

1. A screw compressor can unload down to 10%. Reciprocating compressors are usually limited to:

   a. 10%   **c.** 25%
   b. 15%   **d.** 50%

2. A direct-expansion chiller is designed for 43°F supply chilled water and 57°F return. The freeze stat is set for

   a. 38°F   **c.** 34°F
   b. 36°F   **d.** 32°F

3. A Carrier "Flowtronic Chiller" employs an electronic expansion valve with a superheat setting of:

   a. 12 degrees   **c.** 5 degrees
   b. 10 degrees   **d.** 0 degrees

4. An R-11 system is converted to R-123 with a variable speed drive. To maintain capacity, the converted system will run at:

   a. the same speed
   b. reduced speed
   c. higher speed
   d. an increase in Hertz

5. When converting R-11 to R-123 and polyolester oil:

   a. change mineral oil to polyolester, evacuate the system, and charge and put system back on line
   b. the chiller will function properly with up to 50% mineral oil remaining in the system

   c. Alkylbenzene should be used in place of polyolester oil
   d. Three oil changes are required prior to charging the system with R-123

6. Substituting R-123 for R-11 will:

   a. improve performance
   b. lower performance
   c. not affect performance
   d. increase performance with a higher compression ratio

7. The condensing temperature of a lithium bromide absorption system operating at 100°F is:

   a. 47 psia   **c.** 47 microns
   b. 47mm   **d.** 47 in. Hg

8. In an absorption unit the absorber and generator take the place of the:

   a. compressor   **c.** condenser
   b. evaporator   **d.** receiver

9. Ammonia leaks can be detected with:

   a. a halide torch
   b. an electronic HFC sniffer
   c. bubbles
   d. a sulfur candle

10. The lowest-cost installed thermal storage system is the:

   a. ice storage/refrigerant coils
   b. ice storage/parallel evaporator
   c. compressor-aided partial storage
   d. propylene-glycol cooling-tower system

■ WRITTEN EVALUATION ■

1. List the advantages of a flooded chiller over a direct-expansion chiller.
2. How would you check for leaks with a lithium bromide absorption system?
3. With thermal ice storage what can be gained by utilizing an evaporative condenser over a water-cooling tower?

Unit 16

PSYCHROMETRICS

■ INTRODUCTION

The following tasks will require the use of two psychrometric charts, Figures 16–1 and 16–2.

Unless otherwise specified, the values of the air properties are for atmospheric pressures at sea level: 29.92 mm Hg or 101.325 kPa.

Accurate readings taken from operating equipment can only be obtained if corrections are made for atmospheric conditions other than pressures at sea level, or by using a psychrometric chart that corresponds to atmospheric pressures at higher elevations.

Moreover, plastic charts are available from Carrier that can be conveniently used to plot a cycle and easily erased for reuse.

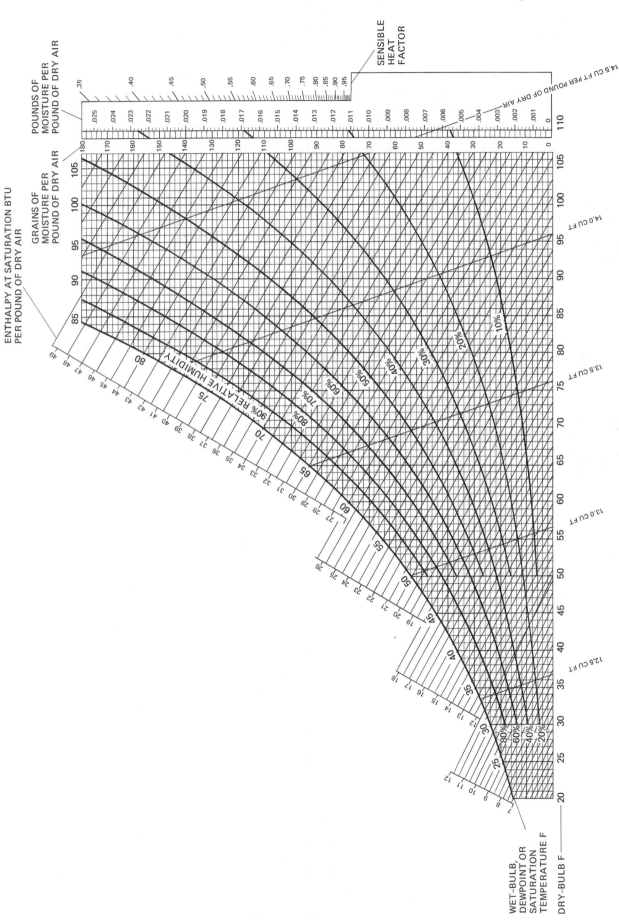

FIGURE 16–1 Psychrometric chart. *(Carrier Corp.)*

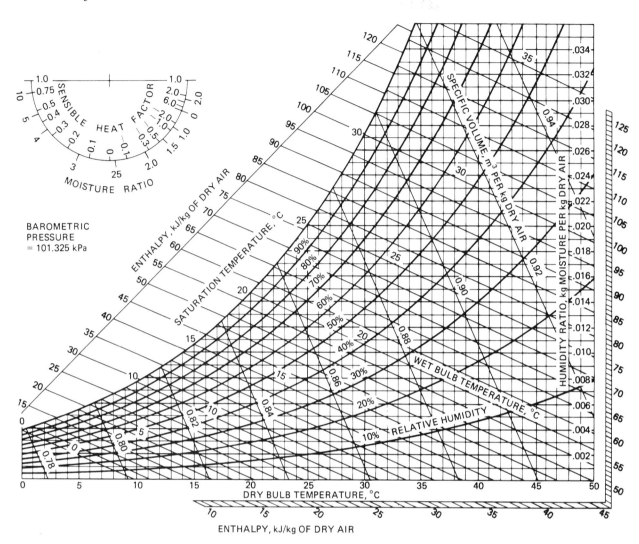

FIGURE 16–2 Psychrometric chart using SI units. *(Business Publishing Co., © 1977.)*

■ JOB SHEET 16–1 ■

Name _____

Score _____ Date _____

PERFORMANCE OBJECTIVE

Given an operating air-conditioning unit, a psychometric chart, and sling psychrometer, complete the following worksheet.

REFERENCE

Heating, Ventilating, and Air Conditioning Fundamentals (Chapter 16); manufacturer's information on British-unit psychrometric chart

EQUIPMENT

Operating air-conditioning unit

TOOLS

Sling psychrometer, ruler, chart (Figure 16–1), pencil, worksheet

JOB 16–1

Completing a psychrometric worksheet

PROCEDURE

1. With a sling psychrometer determine the outside air:

   Wet bulb: _____ Dry bulb: _____
2. Locate state point from step 1 and find the following:
   a. Relative humidity:_____
   b. Specific volume:_____
   c. Total heat:_____
   d. Dewpoint temperature:_____
   e. Moisture content:_____
3. Set thermostat to call for heat and determine the following:
   Dry-bulb temperature of air entering the furnace: _____
   Wet-bulb temperature of air entering the furnace: _____
4. Locate state point of mixed air entering the furnace on the psychrometric chart from temperatures found in step 2.
5. Measure dry-bulb temperature of air leaving furnace and determine final condition of air.
   a. Wet bulb:_____
   b. Quantity of heat added to air:_____
   c. Latent heat added to air:_____
   d. Sensible heat added to air:_____
6. Reset thermostat and allow heat exchanger to cool down to return air temperature.
   Dry-bulb temperature entering air: _____
   Dry-bulb temperature leaving air: _____
7. Reset thermostat to call for cooling.
   a. Locate state point of air entering evaporator coil.
   b. Locate state point of air leaving coil.

8. Draw a line joining entering-air and leaving-air state points and determine the following:
   a. Dewpoint of entering air:_____
   b. Dewpoint of leaving air:_____
   c. Latent heat removed:_____
   d. Sensible heat removed:_____
   e. Sensible heat factor:_____
   f. Specific volume of air leaving the coil:_____
   g. Grains of moisture per pound of dry air removed:

   _____
   h. Relative humidity of air entering coil:_____
   i. Relative humidity of air leaving coil:_____
9. From the dry-bulb temperature of the leaving air, subtract the dewpoint and find:
   Dewpoint depression: _____
10. The difference between leaving-air dry bulb and wet bulb equals:
   Wet-bulb depression:_____

■ JOB SHEET 16–2 ■

Name _____

Score _____ Date _____

PERFORMANCE OBJECTIVE
Given an SI metric psychrometric chart and operable air-conditioning unit, complete the following worksheet.

REFERENCE
Heating, Ventilating, and Air Conditioning Fundamentals (Chapter 16); SI metric chart

EQUIPMENT
Operating air-conditioning unit

TOOLS
Sling psychrometer, ruler, psychrometric chart (Figure 16–2)

JOB 16–2
Completing an SI psychrometric chart worksheet

PROCEDURE
1. Set thermostat to call for cooling and with a sling psychrometer determine the following:
    OSA Dry bulb: _____°C; Wet bulb: _____°C
    Return air dry bulb: _____°C; Wet bulb: _____°C
2. Plot the two state points from step 1 and draw a connecting line.
3. Determine the mixed-air dry-bulb temperature for 25 percent outside air, and 75 percent return air and locate this mixed-air state point on the line drawn in step 2.
4. The mixed-air properties (step 3) are:
    a. Saturation temperature: _____°C
    b. Dewpoint temperature: _____°C
    c. Enthalpy: _____
    d. Humidity ratio: _____
    e. Wet-bulb temperature: _____
    f. Volume of mixture: _____
    g. Relative humidity: _____
5. The wet-bulb temperature of air leaving the coil: _____
6. The dry-bulb temperature of air leaving the coil: _____
7. Locate state points (steps 5 and 6) on chart.
8. Find condensation removal by subtracting humidity ratio from state-point leaving air (step 7) from state-point mixed air (step 3). _____
9. Determine the following:
    Latent heat removal: _____; Sensible heat removal: _____
10. The supply air is set to maintain a dry-bulb temperature of 24°C room thermostat setting. Therefore, the relative humidity of the room will be _____ percent.

■ MULTIPLE-CHOICE TEST ■

Name _____

Score _____ Date _____

DIRECTIONS
Circle the letter that best answers the following multiple-choice questions.

1. Wet-bulb temperature is:

   a. a direct measure of humidity
   b. an effect of moisture content and sensible heat
   c. the percentage of moisture in the air
   d. read by an ordinary thermometer

2. The dry-bulb temperature:

   a. lines are the same as enthalpy lines
   b. lines run parallel to specific volume lines
   c. line meets wet-bulb line at dewpoint
   d. line represents total heat

3. Psychrometrics:

   a. relates to the volume of dry air contained in a pound of moist air
   b. is the thermodynamics of moist air
   c. increases as moisture content is raised
   d. decreases as the total grains of moisture are lowered

4. Enthalpy:

   a. cannot determine the sensible heat factor
   b. is measured in grains per pound of dry air
   c. is measured in cubic feet per pound of moist air
   d. is measured in Btu per pound of air

5. One of the following does not apply:

   a. specific volume relates to air and the space occupied by the air
   b. specific volume is the number of cubic feet occupied by one pound of dry air

   c. specific volume lines run parallel to the dew-point lines
   d. specific volume increases when the wet-bulb temperature increases

6. One of the following does not increase in the heating and humidifying process.

   a. enthalpy
   b. dry-bulb temperature
   c. sensible heat factor
   d. humidity ratio

7. The evaporative cooling process:

   a. raises the relative humidity
   b. lowers the dry-bulb temperature without changing the total heat
   c. converts sensible heat to latent heat
   d. involves all the above

8. The barometric pressure on a psychrometric chart in SI metric is shown in:

   a. J/kg   c. cm^2/kg
   b. kPa   d. N/J^2

9. SI metric psychrometric chart does not show:

   a. grains of moisture
   b. relative humidity
   c. wet-bulb temperature
   d. sensible heat factor

10. SI and British unit psychrometric charts:

    a. use the same temperature reference of absolute zero
    b. use the temperature reference of 0°C and 0°F, respectively
    c. dry-bulb scales at 20° indicate equal enthalpy
    d. use −40 as their datum point

■ WRITTEN EVALUATION ■

1. The equipment designer must know psychrometrics, but how does a basic knowledge of psychrometrics help the service technician?
2. Explain why the total heat for a given wet-bulb and dry-bulb temperature on a psychrometric chart with British units will not be the same if the two temperatures are converted to celsius and the total heat is read from a chart in SI metric.
3. Standard psychrometric charts are prepared for barometric pressures at sea level. Why would altitude corrections be needed for places such as Denver, Colorado or Mexico City?

Unit 17

DUCTWORK

◼ INTRODUCTION

Installation of ductwork is considered the work of the sheet-metal mechanic. However, the current trend has been the increased use of Fiberglas duct. Fabricating and installing a Fiberglas duct system is essentially the same as for sheet metal, except with respect to allowances for the thickness of the board, tools used, and methods of closure.

A Fiberglas duct system can be used for nearly every type of commercial or residential heating, ventilating, or air-conditioning system operating at temperatures to 250°F (121°C) and velocity pressures up to 2,400 ft/min and 2-in static pressure (rigid round duct 5,400 ft/min and 4-in W.G.).

The following performance tasks include duct fabrication and air balancing.

■ JOB SHEET 17–1 ■

Name _____

Score _____ Date _____

PERFORMANCE OBJECTIVE
Given blueprints of the space to be conditioned and necessary supplies, locate on tracing paper the supply and return air registers,
and size ductwork.

REFERENCE
Heating, Ventilating, and Air Conditioning Fundamentals (Chapter 17); manufacturer's register catalog

EQUIPMENT
Drafting machine (optional) and blueprints

TOOLS
Drafting tools

SUPPLIES
Tracing paper

JOB 17–1
Sketching location of registers; sizing necessary ductwork to condition space indicated on blueprint

PROCEDURE
1. Determine proper amount of air to conditioned space.
2. Place tracing paper over blueprint and sketch register locations (perimeter system).
3. Locate fan coil unit.
4. Size supply and return air lines.
5. Evaluate your data with instructor's checklist.

■ JOB SHEET 17–2 ■

Name _____

Score _____ Date _____

PERFORMANCE OBJECTIVE
Given blueprints, air handler, measuring tools, and necessary supplies, lay out and fabricate the duct system or model.

REFERENCE
Heating, Ventilating, and Air Conditioning Fundamentals (Chapter 17); extended plenum systems, *Certain Teed Fibre Glass Fabrication Manual*

EQUIPMENT
Air handler

TOOLS
Shiplap tool, knife, staple gun, yardstick, hand tools

SUPPLIES
Duct board, duct tape, extractors, turning vanes, quadrants, flexible duct, ceiling diffusers, air-balance sheet

JOB 17–2
Fabricating duct system and air balance

PROCEDURE
1. Sketch location of air handler on tracing paper.
2. Locate ceiling outlets and return air grille(s).
3. Locate extended plenum on sketch.
4. Sketch branch lines (flexible duct).
5. Fabricate extended plenum (install turning vanes at 90° ells).
6. Install quadrant flexible-duct tap-ins with extractors.
7. Sketch out, measure, and cut flexible-duct branch lines.
8. Connect flexible duct to ceiling outlets.
9. Balance air as outlined in text.
10. Record air measurements and data on air-balance sheet.
11. Compare air-balance sheet with instructor's findings.

■ MULTIPLE-CHOICE TEST ■

Name _____

Score _____ Date _____

DIRECTIONS:
Circle the letter that best answers the following multiple-choice questions.

1. Dirt-streaked walls or ceilings at air outlets are usually caused by:

   a. dirty filters
   b. dirty coils
   c. dirty ductwork
   d. dirty aspirated air

2. A slope gauge (inclined manometer) with a pitot tube is being used to measure airflow through a trunk line. Cubic feet per minute is determined by multiplying:

   a. free area of duct times total pressure
   b. free area of duct times static pressure
   c. free area of duct times velocity pressure
   d. free area times square root of the velocity pressure times 4005

3. The following hardware can be adjusted at the trunk line to increase the volume of air to a branch outlet:

   a. extractor
   b. directional control
   c. opposed blade damper
   d. turning vanes

4. The total heat gain (latent + sensible) for the master bedroom was calculated at 6000 Btu/h. The minimum air requirement for the room is:

   a. 200 cfm   c. 400 cfm
   b. 300 cfm   d. 450 cfm

5. A power ventilator for a residence is sized at the air-change rate of:

   a. 1 to 2 min   c. 3 to 6 min
   b. 2 to 5 min   d. 5 to 10 min

6. The HUD requirements for proper attic ventilation of a residence can be cut in half when:

   a. the house is erected on a slab with no basement
   b. a vapor barrier is installed in the attic floor

   c. a vapor barrier is installed in the attic ceiling
   d. eave vents are omitted and an oversize gable vent is installed

7. To assist in the even distribution of air through a blow-through steam coil:

   a. expanded metal is installed downstream
   b. expanded metal is installed upstream
   c. expanded metal filter holders are used
   d. a splitter damper directs the airflow across both sides of the coil

8. For a class II, ducted ventilating air handler, the following type of fan would be selected:

   a. propeller fan
   b. forward curved fan
   c. backward inclined fan
   d. air-foil fan

9. One of the following statements is true:

   a. an underfloor plenum system supplies air to perimeter diffusers with flexible duct
   b. an underfloor plenum system has one or more extended plenums
   c. an underfloor plenum system is connected to a box plenum
   d. an underfloor plenum system is constructed with a wood plenum

10. A velometer reading can be interpreted directly as cfm when:

   a. used in conjunction with a flow-measuring hood
   b. the average velocity is squared
   c. the K factor exceeds 0.1 in of water pressure loss per 100 ft of duct
   d. the K factor is multiplied by the free area of the grill

■ PROBLEM ■

1. Size the ductwork for the system (Figure 17–1) using round flexible duct. Reduce trunk-line size only if the size drops 2 inches or more beyond the air outlet. Indicate duct size for each letter (*A* through *J*). (Refer to text Figures 17–26, 17–31, and 17–32.)
2. Indicate rectangular equivalents for each section of round duct (step 1).
3. Calculate friction loss for each section of pipe, *A* through *J*.
4. Determine the external pressure loss *A* through *K* (total pressure loss *A* through *J* plus *K*). Formula for outlet *K*: $(fpm/4005_2$ = velocity pressure loss
5. Branch take off *L* and *M* total 64 ft plus two elbows. The external static is determined by the run with greatest pressure loss. What is the external pressure loss in inches of water from the air handler to outlet number 6?

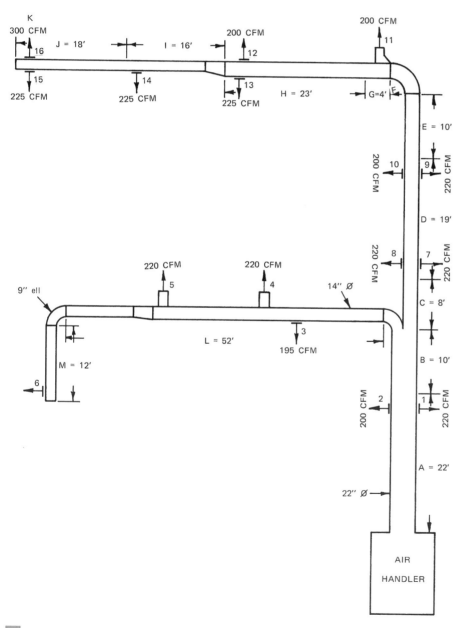

FIGURE 17–1 Ventilating system duct layout.

■ FORMULAS ■

1. $Q = A_e V$
   Q = quantity
   A_e = Net free area (K factor)
   V = velocity

2. $V = 4005 \sqrt{VP}$
   V = velocity
   VP = velocity pressure

3. $VP = \left(\dfrac{\text{ft./min}}{4005} \right)^2$

Example:
Find static pressure loss to move a quantity of air through a register:
400 cfm ÷ 0.45 net free area = 889 ft/min
$VP = (889/4005)^2 = 0.049$ in of water

4. K factor = Q/V
   Q = airflow rate, cfm
   V = averaged measured velocity, ft/min

5. hp required = hp measured

$$\times \left(\dfrac{\text{rpm required}}{\text{rpm measured}} \right)^3$$

Unit 18

STARTUP AND TESTING

■ INTRODUCTION

Various trades are usually involved with the installation and startup of an air-conditioning unitary system: The ductwork is generally installed by a sheet-metal contractor; the unit may be air-balanced by an independent air-balancing company; the unit may be completely wired (including 24-V thermostat wiring) by the electrical contractor; a plumber may run chilled-water, hot-water, and condensate lines, but an air-conditioning service mechanic is needed to evacuate and charge a central air-conditioning system.

The performance tasks to follow are evacuating and charging a central air system. (Refer to Figures 18–1 and 18–2 in the textbook.)

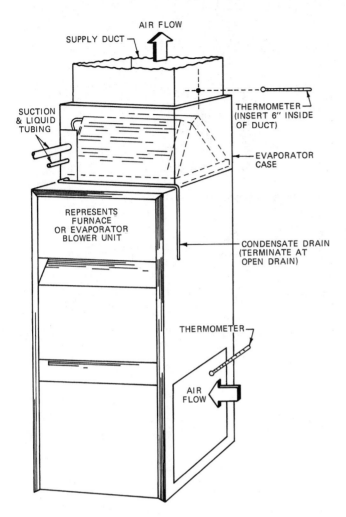

AIR FLOW

SUPPLY DUCT

THERMOMETER
(INSERT 6" INSIDE
OF DUCT)

SUCTION
& LIQUID
TUBING

EVAPORATOR
CASE

REPRESENTS
FURNACE
OR EVAPORATOR
BLOWER UNIT

CONDENSATE DRAIN
(TERMINATE AT
OPEN DRAIN)

THERMOMETER

AIR
FLOW

FIGURE 18–1 Determine air temperature leaving evaporator. *(Southwest Manufacturing— Division of McNeil)*

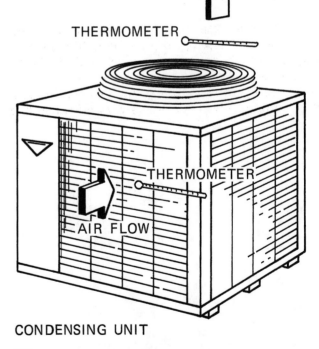

AIR
FLOW

THERMOMETER

THERMOMETER

AIR FLOW

CONDENSING UNIT

FIGURE 18–2 Determine air temperature leaving condensing unit. *(Southwest Manufacturing—Division of McNeil)*

■ JOB SHEET 18–1 ■

Name _____

Score _____ Date _____

PERFORMANCE OBJECTIVE
Given a piped central air-conditioning system, portable high-vacuum charging station, thermistor vacuum gauge, bottle of dry nitrogen, refrigerant, soap-bubble leak detector, and electronic leak detector, evacuate a system free of leaks to 500 μm absolute pressure. The 500 μm should contain less than 200 μm of noncondensible vapor.

REFERENCE
Heating, Ventilating, and Air Conditioning Fundamentals (Chapter 18)

EQUIPMENT
Central air-conditioning installation

TOOLS
Hand tools, portable high-vacuum charging station, thermistor vacuum gauge, electronic leak detector, nitrogen regulator, soap-bubble leak detector

SUPPLIES
Bottle of nitrogen, bottle of refrigerant, shop towels

JOB 18–1
Evacuating the system to within 200 μm of noncondensible vapor

PROCEDURE
1. Remove access service panels from condensing unit and indoor fan coil unit.
2. Clean high- and low-side access valves with shop towel before attempting to connect gauge manifold hoses.
3. Remove valve core from access valves, or midposition suction and discharge service valves. (If system does not have a high- and a low-side service valve, install one on the high and the low side of the system.)
4. Connect the high-pressure manifold gauge to the high-side access valve with the flexible metal line (or red service hose if metal hose is not available).
5. Connect the compound gauge to the low-side access valve with the second flexible metal line (or blue service hose).
6. Close the two manifold valves located in line with the high- and low-side gauges.
7. Close the shut-off valve connecting the vacuum pump to the high-vacuum manifold.
8. Close the shut-off valve connecting the electronic vacuum gauge to the high-vacuum manifold.

9. Open the two shut-off valves located on opposite ends of the white charging hose. This will permit liquid refrigerant to flow from the dial-a-charge cylinder to the gauge manifold.
10. Slowly open the high-side manifold gauge (to permit liquid to flow through the high side of the system).
11. Only 5 or 10 psig (34.47 to 68.94 kPa) are needed.

    NOTE: Find the restriction if the pressure does not build up on the low side (liquid line solenoid if used may have to be manually opened).

12. If pressure is indicated on the low-side gauge, the system is open.
13. Connect nitrogen pressure regulator to nitrogen tank.
14. Close valves on both ends of white charging hose.
15. Disconnect charging hose from dial-a-charge cylinder valve and connect charging hose to nitrogen regulator.
16. Add nitrogen to the system to obtain 50 psig (344.7 kPa).
17. Listen for leaks and use soap bubbles on all weld joints.
18. Observe manifold gauges for loss of pressure.
19. Check system with electronic leak detector if the gauge pressure does not remain constant.
20. If gauge pressure remains constant, pressurize system to 100 psi (689.4 kPa).
21. If system passes step 20, continue to pressurize system to specified maximum operating pressure.
22. Purge system to zero pressure and connect charging hose to dial-a-charge cylinder.
23. Connect electronic vacuum gauge sensor to high-vacuum manifold.
24. Open all service gauges (turn counterclockwise) on high-vacuum manifold and turn on vacuum pump.
25. Evacuate system to 1500 μm.
26. Close vacuum-pump shut-off valve and see if system maintains 1500 μm.

    NOTE: If system builds up to 5000 μm (vapor pressure of ice at 32°F), system has moisture; a higher pressure indicates a leak or vacuum pump has not been run long enough to remove noncondensibles.

27. When the 1500-μm test is passed, close electronic vacuum gauge shut-off valve and open dial-a-charge liquid valve until system vacuum is broken.
28. Open vacuum-pump shut-off valve and electronic vacuum gauge line for second evacuation down to 1500 μm.
29. Break vacuum (step 27).
30. If valve cores were removed from access service valves, pressurize system above atmospheric and reinstall valve cores.
31. Reconnect high- and low-side gauge lines and complete the third evacuation to 500 μm.
32. Charge unit with refrigerant to manufacturer's specifications.

■ JOB SHEET 18–2 ■

Name _____

Score _____ Date _____

PERFORMANCE OBJECTIVE

Given a central air-conditioning unit evacuated and ready to charge, refrigerant, gauge manifold, and hand tools, charge a unit to the manufacturer's specifications. The unit running under load conditions will operate at suction temperatures above freezing. Condensing temperatures will approximate 105°F (40.5°C) for water-cooled units and ambient plus 30°F for air-cooled unit.

REFERENCE

Heating, Ventilating, and Air Conditioning Fundamentals (Chapter 18); nameplate on factory-assembled units; manufacturer's catalogs; dial-a-charge instructions

TOOLS

Gauge manifold, charging station, hand tools, pressure-temperature chart

SUPPLIES

Refrigerant specified on nameplate and matching metering device type

JOB 16–2
Liquid charging unit

PROCEDURE

1. Check installation manual for manufacturer's recommended charging procedures.
2. Locate unit serial plate and note type of refrigerant and amount of charge specified.

   NOTE: Remote systems require additional refrigerant for tubing and evaporator coil.

3. Check metering device to ensure that refrigerant indicated on condensing-unit serial plate matches.
4. Connect charging hose to refrigerant cylinder or dial-a-charge.
5. Purge charging line with refrigerant from charging cylinder.
6. Measure correct amount of liquid into dial-a-charge or weigh refrigerant cylinder with a platform scale.
7. With machine turned off, charge liquid through high-side gauge or liquid-line charging valve to amount specified (step 2).

   NOTE: Large amounts of refrigerant (over 5 lb) can be liquid-charged only while running with the king valve front-seated and the cylinder connected to the gauge port on the king valve or to a separate charging valve installed downstream of the king valve.

8. With the machine running, only refrigerant vapor can be charged into the system, through the low-side manifold port.

9. Close refrigerant cylinder when the low-side gauge reaches a pressure corresponding to 32°F, and observe head pressure if refrigerant is not being weighed while charging the unit.

■ JOB SHEET 18–3 ■

Name _____

Score _____ Date _____

PERFORMANCE OBJECTIVE
Given a central air-conditioning system that is installed and charged with refrigerant, complete the check test and start procedures within a specified period of time. (Results will verify the system is working properly provided that outdoor temperature is 75°F or higher.)

REFERENCE
Heating, Ventilating, and Air Conditioning Fundamentals (Chapter 18); Figures 18–1 and 18–2

EQUIPMENT
Central air-conditioning system

TOOLS
Tape measure, velometer, thermometer, electric drill

SUPPLIES
Unit filters

JOB 18–3
Check testing and starting

PROCEDURE
1. Examine system air filter(s) and replace if dirty or clogged.
2. Examine air-distribution system to determine if volume dampers are open or properly set and all registers are open for free-air delivery.
3. Set thermostat for fan-only operation.
4. With a velometer measure the return air at the evaporator blower unit. A minimum of 400 cfm of air per ton of refrigeration is required.
5. Close the furnace disconnect switch and then close the air-conditioning unit disconnect switch.
6. Drill thermometer access holes as shown in Figure 18–1. Caution: Do not drill a hole in the evaporator.
7. Set cooling thermostat to a setting below room temperature.
8. Allow system to operate approximately 10 minutes and insert pocket thermometers as shown in Figure 18–1. The thermometer reading should indicate a temperature drop of approximately 17 to 19°F dry bulb, below room temperature.
9. While the system is still operating, determine the air temperature rise above outdoor temperature of the air being discharged from the condensing unit (Figure 18–2). The thermometer should indicate a temperature rise of approximately 20° dry bulb above outdoor temperature.

10. Check the evaporator coil when the suction pressure corresponds to a temperature above freezing and the condensing temperature corresponds to ambient plus 30° for air-cooled units or 105°F (40.5°C) for water-cooled units.

11. The temperature drop across the evaporator coil would be approximately 25°.

12. The metering device should have a throttling sound rather than a hissing sound, which indicates a low charge.

13. The coils should sweat but not frost.

14. Amp the compressor motor. The reading should be less than FLA unless the ambient temperature is exceedingly high.

15. Back-seat suction and discharge service valves and remove gauges.

16. Turn off the unit and allow pressures to balance before removing hoses connected to Shraeder (tire valve) type access valves.

17. Replace access panels and service valve caps.

■ MULTIPLE-CHOICE TEST ■

Name _____

Score _____ Date _____

DIRECTIONS

Circle the letter that best answers the following multiple-choice questions.

1. Installation startup is usually performed by:

   a. the apprentice
   b. the sales engineer
   c. the air-balance technician
   d. a competent journeyman

2. The first step in making an electrical inspection upon startup is to:

   a. remove panels and visually check equipment
   b. turn on line voltage disconnect switch
   c. check the current draw of the supply fan
   d. check the compressor for locked rotor and ground

3. Field wiring:

   a. is wired by the equipment manufacturer
   b. is wired by the electrical contractor
   c. is wired by the air-conditioning mechanic
   d. may be wired by the electrical contractor and the air-conditioning mechanic

4. Furnaces with direct-drive blower motors:

   a. do not require a variable pitched pulley
   b. operate at low speed for cooling
   c. operate at high speed for two-stage heating
   d. are involved in all the above

5. The circuit breaker or fused disconnect for a unitary system is sized by:

   a. the full-load compressor current plus the fan current rating
   b. locked rotor compressor current
   c. locked rotor compressor current plus the fan current rating
   d. at least twice the full-load current listed on the serial plate

6. A package unitary system lists on the serial plate 230/208 V, and 208-V power is connected to the unit; therefore:

   a. a larger size circuit breaker or heavier fuses are needed
   b. a higher fan speed is needed for heating
   c. the primary wiring on the low-voltage transformer must be changed
   d. a hard-start kit will probably be needed

7. A propeller-type condenser fanblade is locked onto the motor shaft:

   a. between the condenser and condenser shroud
   b. between the motor and the fan shroud
   c. so that the outer tip of the fanblade centers the shroud opening
   d. is centered between the condenser and fan motor

8. The voltage supplied to a unit must be within what percent of serial plate voltage?

   a. 5 percent   c. 15 percent
   b. 10 percent   d. 20 percent

9. Low vacuums, when triple evacuating, are usually read with:

   a. a compound gauge
   b. a pascal pressure gauge
   c. a mercury manometer
   d. an electronic micron gauge

10. After completing the startup, a sling psychrometer indicates a high wet-bulb temperature. The problem may be corrected by:

    a. turning the motor variable pulley in a few turns
    b. turning it out a few turns
    c. installing a smaller fan pulley
    d. increasing the motor speed

■ WRITTEN EVALUATION ■

1. Before leaving the air-conditioning unit, what needs re-checking?
2. Package heating and air-conditioning systems are usually designed so that a minimum of maintenance is required. However, complete a preventive maintenance checklist for a residential unit.
3. You arrive on the job expecting to find a unitary system, but find an A coil and remote condensing unit installed with the existing furnace. How will the start-up procedures change?

Unit 19
PNEUMATIC CONTROLS

■ INTRODUCTION

The following tasks entail the calibration of Honeywell, Johnson, and Robertshaw receiver controllers. Before tackling these jobs, one must have a basic understanding of pneumatic controls, as outlined in *Heating, Ventilating, and Air Conditioning Fundamentals* (Chapter 19). Receiver-controllers are designed for use with remote temperature, humidity, or pressure transmitters and provide precise control of pneumatic devices.

■ JOB SHEET 19–1A ■

Name _____

Score _____ Date _____

PERFORMANCE OBJECTIVE
With a properly installed controller and main line air pressure (MLP) available, the following steps will provide close calibration. Branch line pressure (BLP) will verify control point.

REFERENCE
Heating, Ventilating, and Air Conditioning Fundamentals (Chapter 19); Honeywell RP920A-D modular pneumatic controller installation instructions

MATERIAL
Honeywell pneumatic calibration kit

JOB 19–1A
Calibration of a Honeywell modular pneumatic controller

PROCEDURE

■ SENSORS

Sensors need no separate calibration, as they are calibrated to the system.

■ SYSTEMS

Factory calibration with field-calculated settings are adequate for startup on most systems. If it is determined that close calibration of the system is required, use the following procedures to calibrate the system:

If CPA is used, apply 9 psi (62 kPa) to Port 9. If remote set point is used, block Port 8.

Use Table 19–1 to convert temperature to either pressure or percentage. To use Table 19–1:

1. Find the correct sensor column.
2. Find the desired temperature in the column.
3. Read equivalent pressure in the far right column.
4. Read equivalent percentage in the far left column.

■ RP920A SYSTEMS

1. Install a temporary receiver gauge (matching primary sensor) or a 0- to 30-psi gauge in the primary sensor gauge port if one is not permanently installed.
2. Apply MLP to system.
3. Install a 0- to 30-psi gauge in BLP test tap (moisten needle before inserting).
4. With sensor at or near the desired control point, adjust the set-point (W1) knob until the BLP equals the center of the controlled device throttling range, e.g., 8 psi.

   NOTE: For most accurate calibration the actual measured variable at the sensor must be ±10 percent of the expected set point.
5. If the sensor is greater than plus or minus 20 percent of the expected set point, remove the sensor tubing from Port 3 and apply a pressure equivalent (expected set point) with a CCT816B Test Set.
6. If set point and primary sensor gauge do not match, remove set point knob and replace it so the set point matches the actual primary sensor gauge reading.
7. Readjust set point (W1) to desired set point.
8. Calibration is complete.

TABLE 19–1   Sensor value to pressure or percentage conversion chart.

OTHER RANGES

TEMPERATURE RANGE (F)					OUTPUT
-25 →125 F	0 – 100 F	40 – 140 F	-40 →160 F	40 – 240 F	PSIG
-25	0	40	-40	40	3.0
-22	2	42	-36	44	3.24
-19	4	44	-32	48	3.48
-16	6	46	-28	52	3.72
-13	8	48	-24	56	3.96
-10	10	50	-20	60	4.2
- 7	12	52	-16	64	4.44
- 4	14	54	-12	68	4.68
- 1	16	56	- 8	72	4.92
2	18	58	- 4	76	5.16
5	20	60	0	80	5.4
8	22	62	4	84	5.64
11	24	64	8	88	5.88
14	26	66	12	92	6.12
17	28	68	16	96	6.36
20	30	70	20	100	6.6
23	32	72	24	104	6.84
26	34	74	28	108	7.08
29	36	76	32	112	7.32
32	38	78	36	116	7.56
35	40	80	40	120	7.8
38	42	82	44	124	8.04
41	44	84	48	128	8.28
44	46	86	52	132	8.52
47	48	88	56	136	8.76
50	50	90	60	140	9.0
53	52	92	64	144	9.24
56	54	94	68	148	9.48
59	56	96	72	152	9.72
62	58	98	76	156	9.96
65	60	100	80	160	10.2
68	62	102	84	164	10.44
71	64	104	88	168	10.68
74	66	106	92	172	10.92
77	68	108	96	176	11.16
80	70	110	100	180	11.4
83	72	112	104	184	11.64
86	74	114	108	188	11.88
89	76	116	112	192	12.12
92	78	118	116	196	12.36
95	80	120	120	200	12.6
98	82	122	124	204	12.84
101	84	124	128	208	13.08
104	86	126	132	212	13.32
107	88	128	136	216	13.56
110	90	130	140	220	13.8
113	92	132	144	224	14.04
116	94	134	148	228	14.28
119	96	136	152	232	14.52
122	98	138	156	236	14.76
125	100	140	160	240	15.0

(Left axis: ACTUAL TEMPERATURE)

40 to 100 F RANGE

ACTUAL TEMPERATURE	OUTPUT PSIG
40	3.0
42	3.4
44	3.8
46	4.2
48	4.6
50	5.0
52	5.4
54	5.8
56	6.2
58	6.6
60	7.0
62	7.4
64	7.8
66	8.2
68	8.6
70	9.0
72	9.4
74	9.8
76	10.2
78	10.6
80	11.0
82	11.4
84	11.8
86	12.2
88	12.6
90	13.0
92	13.4
94	13.8
96	14.2
98	14.6
100	15.0

50 to 90 F RANGE

ACTUAL TEMPERATURE	OUTPUT PSIG
50	3.0
52	3.6
54	4.2
56	4.8
58	5.4
60	6.0
62	6.6
64	7.2
66	7.8
68	8.4
70	9.0
72	9.6
74	10.2
76	10.8
78	11.4
80	12.0
82	12.6
84	13.2
86	13.8
88	14.4
90	15.0

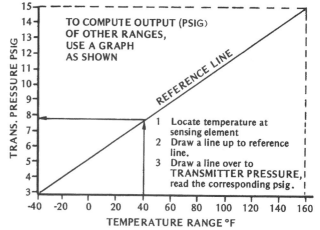

TO COMPUTE OUTPUT (PSIG)
OF OTHER RANGES,
USE A GRAPH
AS SHOWN

REFERENCE LINE

1 Locate temperature at sensing element
2 Draw a line up to reference line.
3 Draw a line over to TRANSMITTER PRESSURE, read the corresponding psig.

(Y-axis: TRANS. PRESSURE PSIG; X-axis: TEMPERATURE RANGE °F)

■ JOB SHEET 19–1B ■

Name _____

Score _____ Date _____

PERFORMANCE OBJECTIVE
With a properly installed controller and main line air pressure (MLP) available, the following steps will provide close calibration. Branch line pressure (BLP) will verify control point.

REFERENCE
Heating, Ventilating, and Air Conditioning Fundamentals (Chapter 19); Honeywell, RP920A-D modular pneumatic controller installation instructions

MATERIAL
Honeywell pneumatic calibration kit

JOB 19–1B
Calibration of a Honeywell modular pneumatic controller

PROCEDURE

■ RP920B SYSTEMS

1. Install a receiver gauge matching the compensation sensor or a 0- to 30-psi gauge to Port 5 as shown in Figure 19–1
   NOTE: If the controller is set up for a compensation gauge in the right gauge port, block Port 5 and use the existing gauge.
2. Install a 0- to 30-psi gage in BLP test tap or in BLP gauge port if one is not permanently installed.
3. Apply MLP to system.
4. Adjust authority knob (A_c) to minimum.
5. Adjust compensation start-point knob (W_c) until receiver gauge reads the pressure equivalent of the compensation start point (Figure 19–2 or 19–3). If W_c and desired start point do not match, remove knob and replace so they do match. Compensation start point is now calibrated.
6. Set up controller calibration tool CCT816B (Figure 19–4) to Port 3 (primary sensor). NOTE: Sensors are not connected.
7. Apply pressure equivalent to the primary sensor value at compensation (comp) start on the reset schedule (Fig. 11 or 12).
8. Adjust the set point (W1) knob until the BLP equals the expected pressure of the controlled device with comp start conditions.

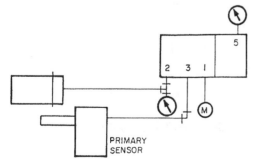

■ **FIGURE 19–1** Piping for RP920B compensation start point setting.

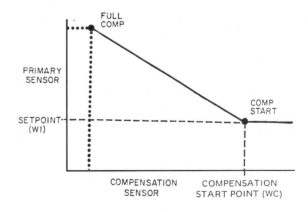

■ **FIGURE 19–2** Negative compensation calibration reset schedule.

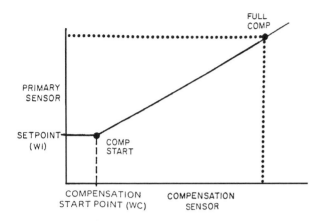

FIGURE 19–3 Positive compensation calibration reset schedule.

9. If needed, remove set-point knob and replace it so the set point matches the primary sensor value at comp start. Set point is now calibrated.
10. Set up a controller calibration tool CCT816B (Figure 19–5) to Port 3 (primary sensor) and Port 5 (compensation sensor).
11. Apply sensor input pressures equivalent to the full compensation on the building schedule.
12. Adjust authority knob (A_c) until the BLP equals the expected pressure of the controlled device with full compensation conditions.
13. Calibration is complete.
14. Changing W_c, W1, or A_c changes the system reset schedule.
15. Remove test set and pipe sensors.

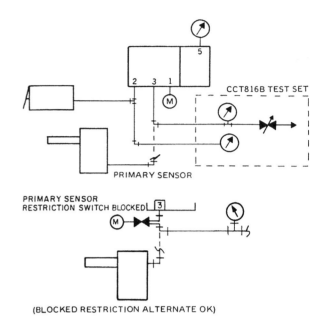

FIGURE 19–4 RP920B set point (W1) test set piping.

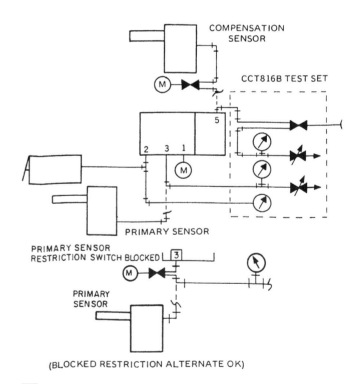

FIGURE 19–5 RP920B authority (A_c) test set piping.

RP920B CALIBRATION EXAMPLE:

Assume the following reset schedule.

Compensation Range	Primary Sensor (Discharge)	Compensation Sensor (OA)
Start	120°F	60°F
Full	160°F	0°F
Operating Span	40°F	60°F
Sensor Range	40 to 240°F	−40 to 160°F

—Assume a 10°F throttling range.
—Use a 2.5 to 6.5 psi operator with an n.o. valve where 6.5 psi is the no-load condition and 2.5 psi is the full load condition.
—See Figure 19–6 for schedule graph.

Initial Settings:

Set point (W1) = 120°F = 40% = 7.8 psi

For Calibration Step 5:
Compensation sensor at start = Compensation start point
$(W_c) = 60°F = 50\% = 9.0$ psi

For Calibration Step 7:
Primary sensor at start = 120°F or 7.8 psi

For Calibration Step 8:
Desired BLP = 6.5 psi

For Calibration Step 12:
Primary sensor at full compensation = 160°F = 10.2 psi
Compensation sensor at full compensation = 0°F = 5.4 psi
Expected BLP = 2.5 psi

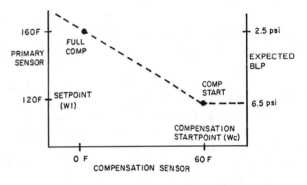

FIGURE 19–6 Calibration example reset schedule.

■ JOB SHEET 19–1C-D ■

PERFORMANCE OBJECTIVE

With a properly installed controller and main line air pressure (MLP) available, the following steps will provide close calibration. Branch line pressure (BLP) will verify control point.

REFERENCE

Heating, Ventilating, and Air Conditioning Fundamentals (Chapter 19); Honeywell, RP920A-D modular pneumatic controller installation instructions

MATERIAL

Honeywell pneumatic calibration kit

JOB 19–1

Calibration of a Honeywell modular pneumatic controller

PROCEDURE

■ RP920C AND D SYSTEMS

1. Remove screw from integral module (Figure 19–7) and install Barb Fitting 14003755-001 in its place.
2. Apply 8.0 psi (55 kPa) to the integral module.
3. Calibrate the RP920C following the procedure for the RP920A.
4. Calibrate the RP920D following the procedure for the RP920B except that the calibration value of the BLP for Steps 8 and 12 should both be 8 psi.
5. When calibration is complete, remove the barb fitting from the integral module and replace the screw.

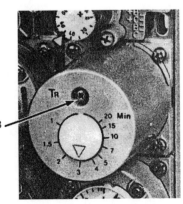

INSTALL BARB FITTING AND APPLY 8.0 PSI (55 KPA)

■ **FIGURE 19–7** Integral module piping.

■ JOB SHEET 19–2 ■

Name _____

Score _____ Date _____

PERFORMANCE OBJECTIVE
Bench calibrate a single-input Robertshaw receiver-controller, with a receiver-controller kit and main air connected to the calibration gauge and to the receiver-controller. Inputs can be adjusted to any value.

REFERENCE
Heating, Ventilating, and Air Conditioning Fundamentals (Chapter 19); Robertshaw Controls Co., Uni-Line Division, *Fundamentals of Pneumatic Controls*

MATERIAL
Robertshaw receiver-controller calibration kit (See Figure 19–8).

JOB 19–2
Bench calibrate a Robertshaw 900-012 receiver-controller

PROCEDURE
1. For this example, use a 0- to 100-degree transmitter.
2. Set the desired throttling range (use 10 degree). 10 divided by 100 is 0.1 or 10 percent. Set TR to 10 percent mark (see Figure 19–9).
3. Set the authority slide to 10 percent or minimum setting.
4. Set the input pressure or reference temperature on the primary input receiver gauge to a value within the range of the transmitter to be used.
5. Adjust the control point adjustment to match the temperature reading by adjusting the branch output pressure to 9 psig (or the midpoint of the spring range) with a calibration tool.
6. Adjust the control-point scale to match the calibration point by lifting the ring and turning it to match input from the primary transmitter. Note that the control-point scale is marked 3 to 15 psig; this corresponds to the 3- to 15-psig output of the primary transmitter.
7. Adjust the control-point adjustment with a calibration tool until the control-point scale indicates the desired set point. This is done by calculating the transmitter output pressure at the desired temperature. Overlays for the control-point scale are available to match the range of the transmitter and can be applied to the control-point scale ring.
8. Once the calibration is made and the control-point scale is set, all adjustments must be made with the set-point adjustment using a calibration tool. If the control-point scale is lifted and turned, the calibration reference is lost.
9. The receiver-controller is now calibrated and can be installed in the system and set to the desired set point.

Now for the calibration procedure utilizing our 900-012 Receiver Controller Calibration Kit. We will utilize this kit because it provides a handy frame of reference for producing inputs to the receiver controller. We refer to this type of calibration as bench calibration, because of the ability to adjust the inputs to any value desired.

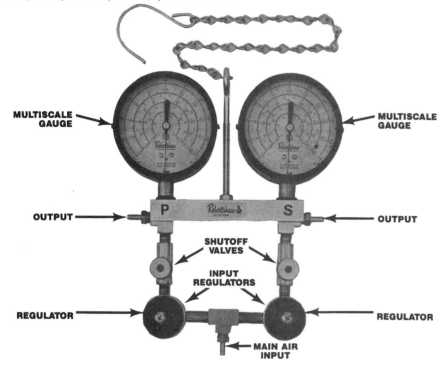

**900-012 RECEIVER CONTROLLER
AND TRANSMITTER CALIBRATION KIT**

The most predominant feature of the receiver controller calibration kit is the multiscale gauges that allow you to read temperatures directly from the gauge scales. The next feature is the pressure regulators used to adjust the output pressures indicated on the gauges. A main air input is provided here, and the simulated transmitter outputs to the receiver controller are attached at the P port for the primary sensor and at S port for the secondary sensor.

FIGURE 19–8   900-012 receiver-controller and transmitter calibration kit. *(Courtesy Robertshaw Controls Co., Uni-Line Division)*

We will be using only the primary signal for the single input calibration. We need to connect main air to our calibration gauges and to the receiver controller, and connect the primary output of the gauge to the primary signal port of the receiver controller.

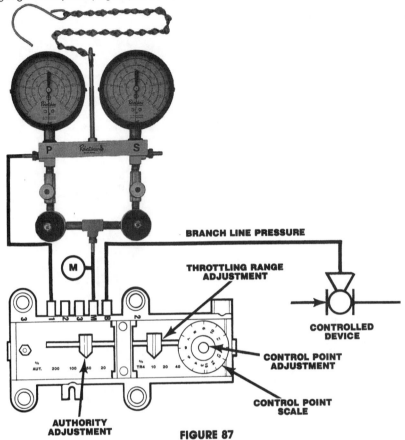

FIGURE 87

The calibration steps for a single input receiver controller are as follows. (For this example we will use the 0° to 100° scale to simulate a 0° to 100° transmitter.)

Set the throttling range to the desired point. We will use a 10° throttling range for this example. 10 ÷ 100 is .10 or 10%, so move the throttling range slide to the 10% mark.

Set the authority slide to 10% or the minimum setting.

Set the input pressure or reference temperature on the primary input receiver gauge to a value within the range of the transmitter to be used.

FIGURE 19–9   Calibration connections.

■ JOB SHEET 19–3A ■

PERFORMANCE OBJECTIVE
Calibrate a Johnson Controls T-5800-1 single-input proportional receiver controller with field-selectable local or remote set point.

REFERENCE
Heating, Ventilating, and Air Conditioning Fundamentals (Chapter 19); Johnson Controls T-5800 installation data

MATERIAL
Johnson X 200-173 calibration kit and X 200-140 test probe with gauge

JOB 19–3A
Calibrate a Johnson single-input receiver-controller

PROCEDURE

■ CALIBRATING THE T-5800-1

See Figure 19–10.

Set-Point Adjustment

To make adjustments to the set point, proceed as follows:

1. Consult the system drawing to determine the required set-point value. Convert this value to the pressure equivalent; see example highlighted in Figure 19–11 (75°F on a 50 to 150°F range would be a set point of 6 psig).
2. Take a set-point pressure reading using the appropriate method and gauge, as illustrated in Figure 19–12. Adjust the local set-point dial or the remote adjuster (manual or automatic) until the pressure matches the value noted in Step 1. Remove the pressure-reading device and reconnect the jumper (if applicable).
3. Start up the system to be controlled. After a reasonable period of time, the receiver-controller should be in control within the throttling range of the controlled device.
4. Proceed to the Gain Adjustment section.

■ GAIN ADJUSTMENT

See Figure 19–13.

Adjusting the gain dial **will not** affect the controller set point; however, the **output pressure** may change when the gain dial is adjusted. Increasing the gain will narrow the throttling range (decrease offset), allowing the control point to be closer to the set point. Decreasing the gain will widen the throttling range, forcing the control point away from the set point.

Normally, having the gain arrow set at the pointer represents a reasonable gain adjustment which would provide stability. Increase the gain setting by small increments until the system becomes unstable and begins to cycle. Decrease the gain setting slightly to remove the cycling effect. Doing so will provide maximum controllability with a minimum of offset.

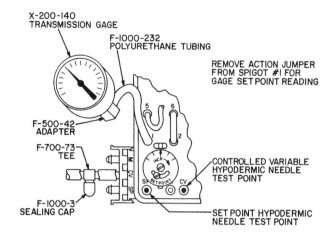

X-200-140
TRANSMISSION GAGE

F-1000-232
POLYURETHANE TUBING

REMOVE ACTION JUMPER
FROM SPIGOT #1 FOR
GAGE SET POINT READING

F-500-42
ADAPTER

F-700-73
TEE

CONTROLLED VARIABLE
HYPODERMIC NEEDLE
TEST POINT

F-1000-3
SEALING CAP

SET POINT HYPODERMIC
NEEDLE TEST POINT

FIGURE 19–10   T-5800-1 adjustment points.

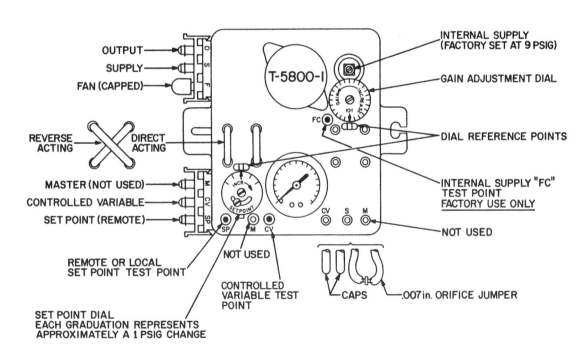

OUTPUT

SUPPLY

FAN (CAPPED)

INTERNAL SUPPLY
(FACTORY SET AT 9 PSIG)

T-5800-I

GAIN ADJUSTMENT DIAL

REVERSE
ACTING

DIRECT
ACTING

DIAL REFERENCE POINTS

MASTER (NOT USED)

CONTROLLED VARIABLE

SET POINT (REMOTE)

INTERNAL SUPPLY "FC"
TEST POINT
FACTORY USE ONLY

NOT USED

REMOTE OR LOCAL
SET POINT TEST POINT

NOT USED

CONTROLLED
VARIABLE TEST
POINT

CAPS

.007 in. ORIFICE JUMPER

SET POINT DIAL
EACH GRADUATION REPRESENTS
APPROXIMATELY A 1 PSIG CHANGE

FIGURE 19–11   Transmitter span vs. output pressure.

TEMPERATURE/PRESSURE/HUMIDITY

0.5 SPAN	25 SPAN		50 SPAN			100 SPAN		200 SPAN	
0.5	65	85	60	80	100	100	150	160	240
.45			55	75	95	90	140	140	220
0.4	60	80	50	70	90	80	130	120	200
.35			45	65	85	70	120	100	180
0.3	55	75	40	60	80	60	110	80	160
.25			35	55	75	50	100	60	140
0.2	50	70	30	50	70	40	90	40	120
.15			25	45	65	30	80	20	100
0.1	45	65	20	40	60	20	70	0	80
.05			15	35	55	10	60	-20	60
0	40	60	10	30	50	0	50	-40	40

(Graph at right: OUTPUT PRESSURE (PSIG), x-axis 3, 5, 6, 7, 9, 11, 13, 15; with reference line at 75 intersecting at 6.)

METRIC CONVERSION FACTORS

(°F – 32) / 1.8 = °C
PSIG x 7 = kPa
in. WG x 249 = Pa

FIGURE 19–12 Methods for checking pressure readings.

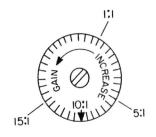

FIGURE 19–13 T-5800-1 gain dial reference points.

■ JOB SHEET 19–3B ■

Name _____

Score _____ Date _____

PERFORMANCE OBJECTIVE
Calibrate a Johnson Controls T-5800-2 single-input proportional re-
ceiver-controller with field-selectable local or remote set point.
This controller provides proportional plus integral control (PI).

REFERENCE
Heating, Ventilating, and Air Conditioning Fundamentals (Chapter 19); Johnson Controls T-5800 installation data

MATERIAL
Johnson X 200-173 calibration kit and X 200-140 test probe with
gauge.

JOB 19–3B
Calibrate a Johnson single-input receiver-controller with proportional integral.

PROCEDURE

■ CALIBRATING THE T-5800-2

See Figure 19–14.

When connecting the T-5800-2 receiver-controller to an operating system, the fan "F" connection should either have the "system in operation"
function signal (minimum of 12 psig) attached
(example: fan on-off or water circulation pump
on-off), or the connection must be capped.

Pulling the P/PI jumper (see Figure 19–15) off of
its spigot causes the receiver-controller to operate
as a proportional-only controller (no integral function). The jumper must be connected in order for
the system to have normal proportional plus integral control.

The T-5800-2 has an automatic/manual integral
control cutout feature when the fan "F" connection is used. This feature keeps the system from
going out of control on startup (after it's been off
for some time) by allowing the system to start up
using proportional-only control. If it is determined that there is not a need for the cutout feature, cap the unused fan "F" connection. Doing so
will allow normal proportional plus integral control of the system at all times, provided that the
P/PI jumper is connected.

■ SET-POINT ADJUSTMENT

To make adjustments to the set point, proceed as
follows:

1. Pull the P/PI jumper off of its spigot so that
   the receiver-controller will operate as a
   proportional-only controller (no integral
   function).
2. Consult the system drawing to determine
   the required set-point value. Convert this
   value to the pressure equivalent; see example highlighted in Figure 19–11 (75°F on a 50
   to 150°F range would be a set point of 6 psig).
3. Take a set-point pressure reading using the
   appropriate method and gauge, as illustrated in Figure 19–12. Adjust the local set-
   point dial or the remote adjuster (manual
   or automatic) until the pressure matches
   the value noted in Step 2. Remove the pressure-reading device and reconnect the
   jumper (if applicable).
4. Start up the system to be controlled.
   After a reasonable period of time, the receiver-controller should be in control (as
   a proportional-only controller) within the
   throttling range of the controlled device.
5. Proceed to the Gain Adjustment section.

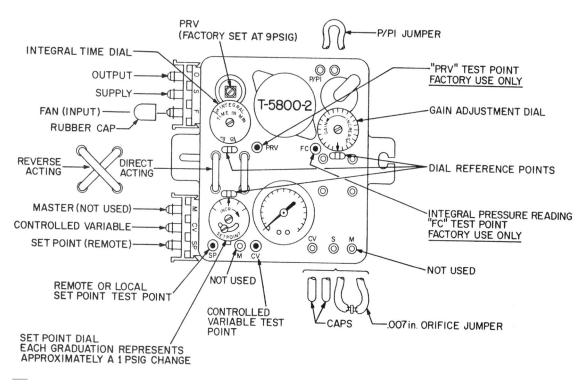

FIGURE 19–14   T-5800-2 adjustment points.

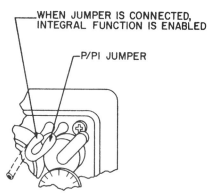

FIGURE 19–15   P/PI jumper located on upper right corner of T-5800-2.

GAIN ADJUSTMENT

See Figure 19–16.

Note: All gain adjustments must be made with the P/PI jumper still removed from the spigot.

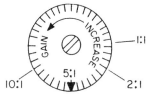

FIGURE 19–16   T-5800-2 gain dial reference points.

Adjusting the gain dial **will not** affect the controller set point; however, the **output pressure** may change when the gain dial is adjusted. Increasing the gain will narrow the throttling range (decrease offset), allowing the control point to be closer to the set point. Decreasing the gain will widen the throttling range, forcing the control point away from the set point.

Normally, having the gain arrow set at the pointer represents a reasonable gain adjustment which would provide stability. Increase the gain setting by small increments until the system becomes unstable and begins to cycle. Decrease the gain setting slightly to remove the cycling effect and mark this position on the dial. Rotate the dial fully clockwise until it hits its mechanical stop, then counterclockwise to the midpoint between the stop and the marked position. Doing so will provide a suitable gain to allow the introduction of the integral function.

After the system stabilizes again, reconnect the P/PI jumper to return the integral function to the receiver-controller. After a reasonable period of time, the control point should stabilize at the set-point value and no further adjustments will be required. If the control point does not stabilize at the set-point value or if excessive cycling occurs, proceed to the Integral Time Adjustment section.

■ INTEGRAL TIME ADJUSTMENT

See Figures 9–16 and 9–17.

Adjusting the integral time dial will not affect the controller set point. If the system response toward the set point is too slow, **decrease** the integral time and/or **increase** the gain-dial settings by small increments each. If cycling occurs, **increase** the integral time and/or **decrease** the gain-dial settings by small increments each.

Mark the set-point dial position. Upset the system by rotating the set-point dial to force the controlled device to an open position. Wait a sufficient period of time to cause controlled variable deviation, then return the set-point dial to its original position. If the system response is not as desired, adjust the integral time and gain dials as prescribed above to obtain the desired system response.

■ SYSTEM START-UP RESPONSE

The following procedure is for determining whether an additional time delay is required for proper system startup.

■ **FIGURE 19–17** T-5800-2 integral time dial reference points.

Warning: Before stopping a system, be sure that a change in output pressure will not upset the system and cause damage.

Receiver-Controllers Using Fan "F" Connection

When the system becomes stable, stop the system and allow enough time to pass until the controlled variable deviates from the set point. (Remember, when the system is off, the receiver-controller returns to proportional-only control. This feature eliminates integral windup during off periods.) Restart the system and observe to see how well it comes into control. When the system is started and the fan signal is increased to maximum (20 psig), the proportional-only controller returns to proportional plus integral control to cause the control point to equal the set point. If the system response is not as desired, determine whether the cause is an incorrect time delay setting (see T-5800-100 Time Delay section) or an incorrect integral time setting (repeat Integral Time Adjustment procedure).

Receiver-Controllers Having Fan "F" Connection Capped

When the system becomes stable, stop the system and disconnect the P/PI jumper. Allow enough time to pass until the controlled variable deviates from the set point. Restart the system, reconnect the P/PI jumper, and observe to see how well the system comes into control. If the system response is not as desired, repeat the Integral Time Adjustment procedure.

■ T-5800-100 TIME DELAY

After system startup, if the system response is too slow, the integral control function will act on the signal before the proportional control function settles out. To eliminate this problem, a T-5800-100 Time Delay (ordered separately) must be added to the fan "F" connection (see Figure 19–18). This device will delay the pressure increase to the fan "F" connection of the receiver-controller, allowing more time for the system to achieve proportional control (in control) before the integral control function is initiated.

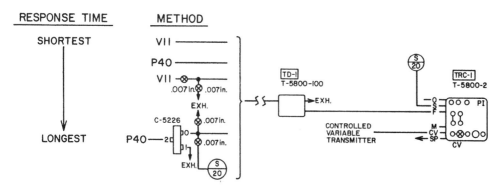

NOTE: INSTALL ONE OF THE ABOVE METHODS
      IF LONGER DELAY TIME IS REQUIRED

FIGURE 19–18   T-5800-100 time delay connected to fan input of T-5800-2.

■ MULTIPLE-CHOICE TEST ■

Name _____
Score _____ Date _____

DIRECTIONS
Circle the letter that best answers the following
multiple-choice questions.

1. Transmitters are:
   a. direct acting
   b. reverse acting
   c. direct or reverse acting
   d. none of the above

2. Receiver-controllers operate at:
   a. 0 to 15 psig   c. 3 to 12 psig
   b. 0 to 20 psig   d. 3 to 15 psig

3. For accurate calibration, the actual measured variable at the sensor must be plus or minus:
   a. 1 percent   c. 5 percent
   b. 2 percent   d. 10 percent

4. With a −40 to 160°F sensor measuring 80°F, the transmitted signal will be:
   a. 120   b. 10.2   c. 9   d. 8

5. After making adjustments to the throttling range, you must:
   a. recalibrate
   b. reset authority
   c. reverse action
   d. set control point scale

6. The BLP calibration point for a receiver-controller is:
   a. 3 psig   c. 12 psig
   b. 8 psig   d. 15 psig

7. Increasing the gain adjustment will :
   a. change set point
   b. widen throttling range
   c. narrow throttling range
   d. change authority

8. To change a Johnson controls T-5800-2 to proportional-only:
   a. pull P / PI jumper off its spigot
   b. turn off the authority
   c. reset the integral control
   d. readjust the derivative spring

9. To readjust gain:
   a. reset the output pressure to 9 psi
   b. change the set point to 8 psi
   c. remove the P/PI jumper from the spigot
   d. calculate the percent authority

10. The required set point is:
   a. matched to ambient conditions
   b. taken from the system drawing
   c. generally 75°F
   d. 70°F

■ WRITTEN EVALUATION ■

1. What happens on startup with a proportional integral controller if the system response is slow? What corrections are needed?
2. How is main air fed to a transmitter?
3. What are some of the advantages of a receiver-controller over the previous master submaster controls?

Unit 20

ENERGY MANAGEMENT SYSTEMS

■ INTRODUCTION

There are a number of building automation control systems on the market. Each one, despite the initial cost, claims to provide a quick payback and a solution to managing a lot of small buildings or a few large buildings. There are many different types of HVAC systems and different types of control systems. The following tasks utilize the TEC/CUBE Energy Management Training System. The objectives are to train the student to understand the basic concepts and principles of the various methods of managing energy.

Tasks include connecting the computer to energy management hardware, entering data to accomplish the desired energy management operation, and training the student to troubleshoot by isolating the various devices in an energy management system. (See Figure 20–1.)

In place of taking a multiple-choice quiz at the end of this unit, you will be required to develop an input/output summary (points matrix) for a multizone unit. For effective energy savings throughout the building, some points are started or stopped (turned on or off), and some are adjusted (resets, DDC). Other points are for analysis, energy management, and/or recordkeeping.

Selecting input/output summary data is extremely important for two reasons:

1. Return on investment is usually a major factor for selecting an energy management system. Some points provide a high return and others are marginal.
2. More than one-half the total system cost is often invested in hardware and wiring remote from the DDC controllers.

PDS113-11/91

CSI's Model 7440 Distributed Control Unit (DCU) is an economical and powerful microprocessor based controller used in the S7000 Building Automation System. The 7440 DCU is designed to connect directly to the controller LAN (Local Area Network) and works in conjunction with all other controllers and workstations on the LAN. The 7440 DCU is capable of monitoring and controlling over 300 addressable points while providing full Direct Digital Control (DDC), energy management functions, and process control in a single complete package.

CONTROLLER DESIGN

The 7440 DCU hardware is built around a single board computer which combines processing, memory, communications and field Input/Output (I/O) functions on a single printed circuit board. All hardwired I/O, power and communication connections to the controller feature quick disconnect terminals to eliminate the need to disconnect individual wires if a controller needs to be changed. The inherent reliability of this monolithic design is further enhanced with the extensive transient protection and automatic self-test features.

DCU FEATURES

Direct Digital Control

- Exclusive modular approach in linking DDC modules
- Easy to tune Proportional Integral Derivative (PID) and Floating Module Control
- Pulse Width Modulation (PWM) or Analog Output (A/O)

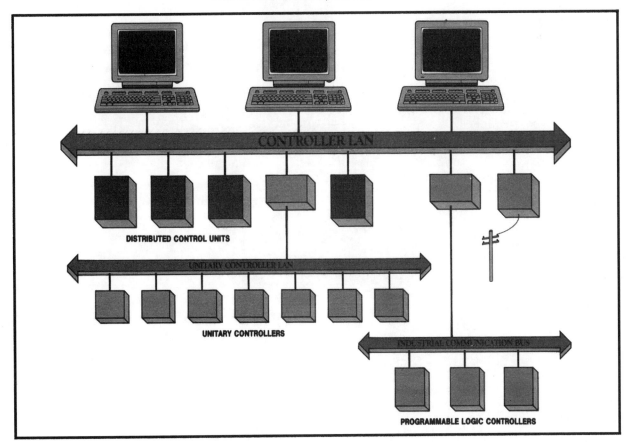

FIGURE 20–1   Model 7440 distributed control unit. *(Control systems International)*

■ JOB SHEET 20–1 ■

PERFORMANCE OBJECTIVE

Train the student to wire control systems to operate per a sequence of operation using the concepts and principles of managing energy.

REFERENCE

Heating, Ventilating, and Air-Conditioning Fundamentals (Chapter 20); TEC/CUBE Energy Management Training System, program manual (Figure 20–1)

SUPPLIES

TEC/CUBE Interface,computer and connecting wiring

JOB 20–1

PROCEDURE

■ SETUP

See Figure 20–2.

CAUTION: This equipment is extremely sensitive to rough handling!

1. Place the computer next to power base and plug the power cord into either a wall socket or the back of the power base. Use a spike protector to prevent damage to the computer from power surges.
2. Place the monitor (T.V.) on the computer, plug gray cord into the back of the computer (there is only one socket that it fits) and plug the black cord into a standard wall outlet.
3. Place the keyboard in front of the monitor and plug it into the back of the computer.
4. Remove the cardboard inserts from the disk drives on the front of the computer, then insert the program disk *A* into drive *A* and the data disk *B* into drive *B*. Pull down the levers on both drives.
5. Set the interface cube on the power base and connect the two gray cables from the computer on the back of the interface cube and screw them into place.
6. Turn on the monitor (rocker switch in the rear) and the computer. The monitor should be doing something and the red light on drive *A* will come on. If everything was done right, the copyright screens will appear. Then the main menu will appear and wait for some keyboard input. If you get this far and the program does not start running on the monitor screen, go back over the previous steps.
7. If the interface is not functioning properly, make sure the cables are connected to the right ports on the back of the computer.

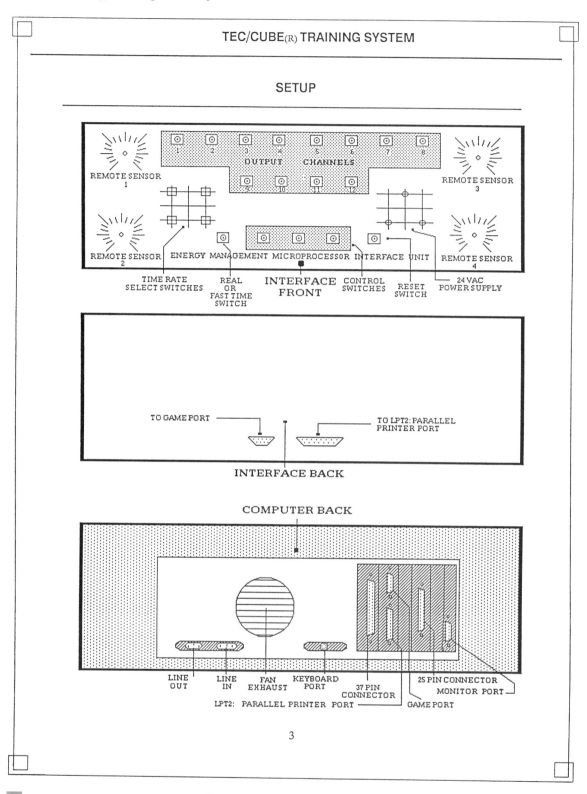

FIGURE 20–2   TEC/CUBE^R training system setup.

■ PROCEDURE

1. Complete the above set-up procedures.
2. Mount 1- to 12-line voltage or low-voltage simulators on the power base (use multiple power bases if necessary).
3. Mount controllers on top of the line-voltage simulators (starters or contactors).
4. Wire from the gray ("T") side of the 24-VAC power supply on the front of the interface unit to each of the gray terminals on the low-voltage simulators and controllers.
5. Wire from each 24-VAC output channel (1–12) to the brown terminal of each low-voltage simulator or controller (any other external controls such as high and low pressure or overload would be wired in between the output channel (1–12) and the device being controlled by that channel).
6. The "V" (brown) terminal on the front of the interface unit is *only* used in conjunction with the TEC/CUBE model EC/085 manual by-pass timer. If you wish to use the by-pass timer, wire from the "V" terminal of the interface unit to one side of the by-pass timer. Then wire from the other side of the by-pass timer to the output channel you wish to override.

■ JOB SHEET 20–2 ■

Name _____

Score _____ Date _____

PERFORMANCE OBJECTIVE
Train the student to enter the necessary data to accomplish the desired energy management operation.

REFERENCE
Heating, Ventilating, and Air Conditioning Fundamentals (Chapter 20); TEC/CUBE program manual

SUPPLIES
Copies of TEC/CUBE schedules and pencil

JOB 20–2
Remove forms from the workbook, make copies, and complete schedules before entering data in the computer.

PROCEDURE
1. Fill in Table 20–1 (Automatic Time Control Schedule).
2. Fill in Table 20–2 (Automatic Temperature Control Schedule).
3. Fill in Table 20–3 (Time And Temperature Control Schedule).
4. Fill in Table 20–4 (Duty Cycle Schedule).
5. Fill in Table 20–5 (Load Shedding Schedule).
6. Select Reset Boiler Control Schedule (Table 20–6).
7. Select Reset Cooler Control Schedule (Table 20–7).

TABLE 20–1   Energy management training unit.

TEC/CUBE(R) TRAINING SYSTEM

APPENDIX C

The forms on this and the following pages are for making schedules before entering them into the computer. We suggest removing them from the manual and using copies.

AUTMATIC TIME CONTROL SCHEDULE

	DAY 1	DAY 2	DAY 3	DAY 4	DAY 5	DAY 6	DAY 7
CHANNEL 1	ON ___ OFF ___ ON ___ OFF ___ ON ___ OFF ___ ON ___ OFF ___	ON ___ OFF ___ ON ___ OFF ___ ON ___ OFF ___ ON ___ OFF ___	ON ___ OFF ___ ON ___ OFF ___ ON ___ OFF ___ ON ___ OFF ___	ON ___ OFF ___ ON ___ OFF ___ ON ___ OFF ___ ON ___ OFF ___	ON ___ OFF ___ ON ___ OFF ___ ON ___ OFF ___ ON ___ OFF ___	ON ___ OFF ___ ON ___ OFF ___ ON ___ OFF ___ ON ___ OFF ___	ON ___ OFF ___ ON ___ OFF ___ ON ___ OFF ___ ON ___ OFF ___
CHANNEL 2	ON ___ OFF ___ ON ___ OFF ___ ON ___ OFF ___ ON ___ OFF ___	ON ___ OFF ___ ON ___ OFF ___ ON ___ OFF ___ ON ___ OFF ___	ON ___ OFF ___ ON ___ OFF ___ ON ___ OFF ___ ON ___ OFF ___	ON ___ OFF ___ ON ___ OFF ___ ON ___ OFF ___ ON ___ OFF ___	ON ___ OFF ___ ON ___ OFF ___ ON ___ OFF ___ ON ___ OFF ___	ON ___ OFF ___ ON ___ OFF ___ ON ___ OFF ___ ON ___ OFF ___	ON ___ OFF ___ ON ___ OFF ___ ON ___ OFF ___ ON ___ OFF ___
CHANNEL 3	ON ___ OFF ___ ON ___ OFF ___ ON ___ OFF ___ ON ___ OFF ___	ON ___ OFF ___ ON ___ OFF ___ ON ___ OFF ___ ON ___ OFF ___	ON ___ OFF ___ ON ___ OFF ___ ON ___ OFF ___ ON ___ OFF ___	ON ___ OFF ___ ON ___ OFF ___ ON ___ OFF ___ ON ___ OFF ___	ON ___ OFF ___ ON ___ OFF ___ ON ___ OFF ___ ON ___ OFF ___	ON ___ OFF ___ ON ___ OFF ___ ON ___ OFF ___ ON ___ OFF ___	ON ___ OFF ___ ON ___ OFF ___ ON ___ OFF ___ ON ___ OFF ___
CHANNEL 4	ON ___ OFF ___ ON ___ OFF ___ ON ___ OFF ___ ON ___ OFF ___	ON ___ OFF ___ ON ___ OFF ___ ON ___ OFF ___ ON ___ OFF ___	ON ___ OFF ___ ON ___ OFF ___ ON ___ OFF ___ ON ___ OFF ___	ON ___ OFF ___ ON ___ OFF ___ ON ___ OFF ___ ON ___ OFF ___	ON ___ OFF ___ ON ___ OFF ___ ON ___ OFF ___ ON ___ OFF ___	ON ___ OFF ___ ON ___ OFF ___ ON ___ OFF ___ ON ___ OFF ___	ON ___ OFF ___ ON ___ OFF ___ ON ___ OFF ___ ON ___ OFF ___
CHANNEL 5	ON ___ OFF ___ ON ___ OFF ___ ON ___ OFF ___ ON ___ OFF ___	ON ___ OFF ___ ON ___ OFF ___ ON ___ OFF ___ ON ___ OFF ___	ON ___ OFF ___ ON ___ OFF ___ ON ___ OFF ___ ON ___ OFF ___	ON ___ OFF ___ ON ___ OFF ___ ON ___ OFF ___ ON ___ OFF ___	ON ___ OFF ___ ON ___ OFF ___ ON ___ OFF ___ ON ___ OFF ___	ON ___ OFF ___ ON ___ OFF ___ ON ___ OFF ___ ON ___ OFF ___	ON ___ OFF ___ ON ___ OFF ___ ON ___ OFF ___ ON ___ OFF ___
CHANNEL 6	ON ___ OFF ___ ON ___ OFF ___ ON ___ OFF ___ ON ___ OFF ___	ON ___ OFF ___ ON ___ OFF ___ ON ___ OFF ___ ON ___ OFF ___	ON ___ OFF ___ ON ___ OFF ___ ON ___ OFF ___ ON ___ OFF ___	ON ___ OFF ___ ON ___ OFF ___ ON ___ OFF ___ ON ___ OFF ___	ON ___ OFF ___ ON ___ OFF ___ ON ___ OFF ___ ON ___ OFF ___	ON ___ OFF ___ ON ___ OFF ___ ON ___ OFF ___ ON ___ OFF ___	ON ___ OFF ___ ON ___ OFF ___ ON ___ OFF ___ ON ___ OFF ___

TABLE 20–1 (Continued)

TEC/CUBE(R) TRAINING SYSTEM

AUTMATIC TIME CONTROL SCHEDULE

	DAY 1	DAY 2	DAY 3	DAY 4	DAY 5	DAY 6	DAY 7
CHANNEL 7	ON ____ OFF ____ ON ____ OFF ____ ON ____ OFF ____ ON ____ OFF ____	ON ____ OFF ____ ON ____ OFF ____ ON ____ OFF ____ ON ____ OFF ____	ON ____ OFF ____ ON ____ OFF ____ ON ____ OFF ____ ON ____ OFF ____	ON ____ OFF ____ ON ____ OFF ____ ON ____ OFF ____ ON ____ OFF ____	ON ____ OFF ____ ON ____ OFF ____ ON ____ OFF ____ ON ____ OFF ____	ON ____ OFF ____ ON ____ OFF ____ ON ____ OFF ____ ON ____ OFF ____	ON ____ OFF ____ ON ____ OFF ____ ON ____ OFF ____ ON ____ OFF ____
CHANNEL 8	ON ____ OFF ____ ON ____ OFF ____ ON ____ OFF ____ ON ____ OFF ____	ON ____ OFF ____ ON ____ OFF ____ ON ____ OFF ____ ON ____ OFF ____	ON ____ OFF ____ ON ____ OFF ____ ON ____ OFF ____ ON ____ OFF ____	ON ____ OFF ____ ON ____ OFF ____ ON ____ OFF ____ ON ____ OFF ____	ON ____ OFF ____ ON ____ OFF ____ ON ____ OFF ____ ON ____ OFF ____	ON ____ OFF ____ ON ____ OFF ____ ON ____ OFF ____ ON ____ OFF ____	ON ____ OFF ____ ON ____ OFF ____ ON ____ OFF ____ ON ____ OFF ____
CHANNEL 9	ON ____ OFF ____ ON ____ OFF ____ ON ____ OFF ____ ON ____ OFF ____	ON ____ OFF ____ ON ____ OFF ____ ON ____ OFF ____ ON ____ OFF ____	ON ____ OFF ____ ON ____ OFF ____ ON ____ OFF ____ ON ____ OFF ____	ON ____ OFF ____ ON ____ OFF ____ ON ____ OFF ____ ON ____ OFF ____	ON ____ OFF ____ ON ____ OFF ____ ON ____ OFF ____ ON ____ OFF ____	ON ____ OFF ____ ON ____ OFF ____ ON ____ OFF ____ ON ____ OFF ____	ON ____ OFF ____ ON ____ OFF ____ ON ____ OFF ____ ON ____ OFF ____
CHANNEL 10	ON ____ OFF ____ ON ____ OFF ____ ON ____ OFF ____ ON ____ OFF ____	ON ____ OFF ____ ON ____ OFF ____ ON ____ OFF ____ ON ____ OFF ____	ON ____ OFF ____ ON ____ OFF ____ ON ____ OFF ____ ON ____ OFF ____	ON ____ OFF ____ ON ____ OFF ____ ON ____ OFF ____ ON ____ OFF ____	ON ____ OFF ____ ON ____ OFF ____ ON ____ OFF ____ ON ____ OFF ____	ON ____ OFF ____ ON ____ OFF ____ ON ____ OFF ____ ON ____ OFF ____	ON ____ OFF ____ ON ____ OFF ____ ON ____ OFF ____ ON ____ OFF ____
CHANNEL 11	ON ____ OFF ____ ON ____ OFF ____ ON ____ OFF ____ ON ____ OFF ____	ON ____ OFF ____ ON ____ OFF ____ ON ____ OFF ____ ON ____ OFF ____	ON ____ OFF ____ ON ____ OFF ____ ON ____ OFF ____ ON ____ OFF ____	ON ____ OFF ____ ON ____ OFF ____ ON ____ OFF ____ ON ____ OFF ____	ON ____ OFF ____ ON ____ OFF ____ ON ____ OFF ____ ON ____ OFF ____	ON ____ OFF ____ ON ____ OFF ____ ON ____ OFF ____ ON ____ OFF ____	ON ____ OFF ____ ON ____ OFF ____ ON ____ OFF ____ ON ____ OFF ____
CHANNEL 12	ON ____ OFF ____ ON ____ OFF ____ ON ____ OFF ____ ON ____ OFF ____	ON ____ OFF ____ ON ____ OFF ____ ON ____ OFF ____ ON ____ OFF ____	ON ____ OFF ____ ON ____ OFF ____ ON ____ OFF ____ ON ____ OFF ____	ON ____ OFF ____ ON ____ OFF ____ ON ____ OFF ____ ON ____ OFF ____	ON ____ OFF ____ ON ____ OFF ____ ON ____ OFF ____ ON ____ OFF ____	ON ____ OFF ____ ON ____ OFF ____ ON ____ OFF ____ ON ____ OFF ____	ON ____ OFF ____ ON ____ OFF ____ ON ____ OFF ____ ON ____ OFF ____

TABLE 20–2   Setup for energy management trainer.

TEC/CUBE(R) TRAINING SYSTEM

AUTOMATIC TEMPERATURE CONTROL SCHEDULE

UPPER SET POINTS

SENSOR 1	SET POINT __ O. C. _____
SENSOR 2	SET POINT __ O. C. _____
SENSOR 3	SET POINT __ O. C. _____
SENSOR 4	SET POINT __ O. C. _____

LOWER SET POINTS

SENSOR 1	SET POINT __ O. C. _____
SENSOR 2	SET POINT __ O. C. _____
SENSOR 3	SET POINT __ O. C. _____
SENSOR 4	SET POINT __ O. C. _____

TABLE 20-3 Input/output points summary.

TEC/CUBE(R) TRAINING SYSTEM

AUTOMATIC TIME/TEMPERATURE CONTROL SCHEDULE

	DAY 1	DAY 2	DAY 3	DAY 4	DAY 5	DAY 6	DAY 7
SENSOR 1	OC.____ UC.____ OC.____ UC.____ OC.____ UC.____ OC.____ UC.____	OC.____ UC.____ OC.____ UC.____ OC.____ UC.____ OC.____ UC.____	OC.____ UC.____ OC.____ UC.____ OC.____ UC.____ OC.____ UC.____	OC.____ UC.____ OC.____ UC.____ OC.____ UC.____ OC.____ UC.____	OC.____ UC.____ OC.____ UC.____ OC.____ UC.____ OC.____ UC.____	OC.____ UC.____ OC.____ UC.____ OC.____ UC.____ OC.____ UC.____	OC.____ UC.____ OC.____ UC.____ OC.____ UC.____ OC.____ UC.____
	O.H.S.P.__ U.H.S.P.__ O.CH. __ O.L.S.P. __ U.L.S.P. __ O.CH. __ O.W.I.S. __ O.W.M. __	O.H.S.P.__ U.H.S.P.__ O.CH. __ O.L.S.P. __ U.L.S.P. __ O.CH. __ O.W.I.S. __ O.W.M. __	O.H.S.P.__ U.H.S.P.__ O.CH. __ O.L.S.P. __ U.L.S.P. __ O.CH. __ O.W.I.S. __ O.W.M. __	O.H.S.P.__ U.H.S.P.__ O.CH. __ O.L.S.P. __ U.L.S.P. __ O.CH. __ O.W.I.S. __ O.W.M. __	O.H.S.P.__ U.H.S.P.__ O.CH. __ O.L.S.P. __ U.L.S.P. __ O.CH. __ O.W.I.S. __ O.W.M. __	O.H.S.P.__ U.H.S.P.__ O.CH. __ O.L.S.P. __ U.L.S.P. __ O.CH. __ O.W.I.S. __ O.W.M. __	O.H.S.P.__ U.H.S.P.__ O.CH. __ O.L.S.P. __ U.L.S.P. __ O.CH. __ O.W.I.S. __ O.W.M. __
SENSOR 2	OC.____ UC.____ OC.____ UC.____ OC.____ UC.____ OC.____ UC.____	OC.____ UC.____ OC.____ UC.____ OC.____ UC.____ OC.____ UC.____	OC.____ UC.____ OC.____ UC.____ OC.____ UC.____ OC.____ UC.____	OC.____ UC.____ OC.____ UC.____ OC.____ UC.____ OC.____ UC.____	OC.____ UC.____ OC.____ UC.____ OC.____ UC.____ OC.____ UC.____	OC.____ UC.____ OC.____ UC.____ OC.____ UC.____ OC.____ UC.____	OC.____ UC.____ OC.____ UC.____ OC.____ UC.____ OC.____ UC.____
	O.H.S.P.__ U.H.S.P.__ O.CH. __ O.L.S.P. __ U.L.S.P. __ O.CH. __ O.W.I.S. __ O.W.M. __	O.H.S.P.__ U.H.S.P.__ O.CH. __ O.L.S.P. __ U.L.S.P. __ O.CH. __ O.W.I.S. __ O.W.M. __	O.H.S.P.__ U.H.S.P.__ O.CH. __ O.L.S.P. __ U.L.S.P. __ O.CH. __ O.W.I.S. __ O.W.M. __	O.H.S.P.__ U.H.S.P.__ O.CH. __ O.L.S.P. __ U.L.S.P. __ O.CH. __ O.W.I.S. __ O.W.M. __	O.H.S.P.__ U.H.S.P.__ O.CH. __ O.L.S.P. __ U.L.S.P. __ O.CH. __ O.W.I.S. __ O.W.M. __	O.H.S.P.__ U.H.S.P.__ O.CH. __ O.L.S.P. __ U.L.S.P. __ O.CH. __ O.W.I.S. __ O.W.M. __	O.H.S.P.__ U.H.S.P.__ O.CH. __ O.L.S.P. __ U.L.S.P. __ O.CH. __ O.W.I.S. __ O.W.M. __
SENSOR 3	OC.____ UC.____ OC.____ UC.____ OC.____ UC.____ OC.____ UC.____	OC.____ UC.____ OC.____ UC.____ OC.____ UC.____ OC.____ UC.____	OC.____ UC.____ OC.____ UC.____ OC.____ UC.____ OC.____ UC.____	OC.____ UC.____ OC.____ UC.____ OC.____ UC.____ OC.____ UC.____	OC.____ UC.____ OC.____ UC.____ OC.____ UC.____ OC.____ UC.____	OC.____ UC.____ OC.____ UC.____ OC.____ UC.____ OC.____ UC.____	OC.____ UC.____ OC.____ UC.____ OC.____ UC.____ OC.____ UC.____
	O.H.S.P.__ U.H.S.P.__ O.CH. __ O.L.S.P. __ U.L.S.P. __ O.CH. __ O.W.I.S. __ O.W.M. __	O.H.S.P.__ U.H.S.P.__ O.CH. __ O.L.S.P. __ U.L.S.P. __ O.CH. __ O.W.I.S. __ O.W.M. __	O.H.S.P.__ U.H.S.P.__ O.CH. __ O.L.S.P. __ U.L.S.P. __ O.CH. __ O.W.I.S. __ O.W.M. __	O.H.S.P.__ U.H.S.P.__ O.CH. __ O.L.S.P. __ U.L.S.P. __ O.CH. __ O.W.I.S. __ O.W.M. __	O.H.S.P.__ U.H.S.P.__ O.CH. __ O.L.S.P. __ U.L.S.P. __ O.CH. __ O.W.I.S. __ O.W.M. __	O.H.S.P.__ U.H.S.P.__ O.CH. __ O.L.S.P. __ U.L.S.P. __ O.CH. __ O.W.I.S. __ O.W.M. __	O.H.S.P.__ U.H.S.P.__ O.CH. __ O.L.S.P. __ U.L.S.P. __ O.CH. __ O.W.I.S. __ O.W.M. __
SENSOR 4	OC.____ UC.____ OC.____ UC.____ OC.____ UC.____ OC.____ UC.____	OC.____ UC.____ OC.____ UC.____ OC.____ UC.____ OC.____ UC.____	OC.____ UC.____ OC.____ UC.____ OC.____ UC.____ OC.____ UC.____	OC.____ UC.____ OC.____ UC.____ OC.____ UC.____ OC.____ UC.____	OC.____ UC.____ OC.____ UC.____ OC.____ UC.____ OC.____ UC.____	OC.____ UC.____ OC.____ UC.____ OC.____ UC.____ OC.____ UC.____	OC.____ UC.____ OC.____ UC.____ OC.____ UC.____ OC.____ UC.____
	O.H.S.P.__ U.H.S.P.__ O.CH. __ O.L.S.P. __ U.L.S.P. __ O.CH. __ O.W.I.S. __ O.W.M. __	O.H.S.P.__ U.H.S.P.__ O.CH. __ O.L.S.P. __ U.L.S.P. __ O.CH. __ O.W.I.S. __ O.W.M. __	O.H.S.P.__ U.H.S.P.__ O.CH. __ O.L.S.P. __ U.L.S.P. __ O.CH. __ O.W.I.S. __ O.W.M. __	O.H.S.P.__ U.H.S.P.__ O.CH. __ O.L.S.P. __ U.L.S.P. __ O.CH. __ O.W.I.S. __ O.W.M. __	O.H.S.P.__ U.H.S.P.__ O.CH. __ O.L.S.P. __ U.L.S.P. __ O.CH. __ O.W.I.S. __ O.W.M. __	O.H.S.P.__ U.H.S.P.__ O.CH. __ O.L.S.P. __ U.L.S.P. __ O.CH. __ O.W.I.S. __ O.W.M. __	O.H.S.P.__ U.H.S.P.__ O.CH. __ O.L.S.P. __ U.L.S.P. __ O.CH. __ O.W.I.S. __ O.W.M. __

OC. = OCCUPIED
UC. = UNOCCUPIED

O.H.S.P. = OCCUPIED HIGH SET POINT
U.H.S.P. = UNOCCUPIED HIGH SET POINT
O.L.S.P. = OCCUPIED LOW SET POINT
U.L.S.P. = UNOCCUPIED LOW SET POINT

O.CH. = OUTPUT CHANNEL
O.W.I.S. = OPT. WARMUP INPUT SENSOR
O.W.M. = OPT. WARMUP MINUTES

TABLE 20–4 Duty cycle schedule.

CHANNEL 1	OFF CYCLE STARTS AT ___	OFF CYCLE LENGTH ___	REPEAT ___ TIMES
CHANNEL 2	OFF CYCLE STARTS AT ___	OFF CYCLE LENGTH ___	REPEAT ___ TIMES
CHANNEL 3	OFF CYCLE STARTS AT ___	OFF CYCLE LENGTH ___	REPEAT ___ TIMES
CHANNEL 4	OFF CYCLE STARTS AT ___	OFF CYCLE LENGTH ___	REPEAT ___ TIMES
CHANNEL 5	OFF CYCLE STARTS AT ___	OFF CYCLE LENGTH ___	REPEAT ___ TIMES
CHANNEL 6	OFF CYCLE STARTS AT ___	OFF CYCLE LENGTH ___	REPEAT ___ TIMES
CHANNEL 7	OFF CYCLE STARTS AT ___	OFF CYCLE LENGTH ___	REPEAT ___ TIMES
CHANNEL 8	OFF CYCLE STARTS AT ___	OFF CYCLE LENGTH ___	REPEAT ___ TIMES
CHANNEL 9	OFF CYCLE STARTS AT ___	OFF CYCLE LENGTH ___	REPEAT ___ TIMES
CHANNEL 10	OFF CYCLE STARTS AT ___	OFF CYCLE LENGTH ___	REPEAT ___ TIMES
CHANNEL 11	OFF CYCLE STARTS AT ___	OFF CYCLE LENGTH ___	REPEAT ___ TIMES
CHANNEL 12	OFF CYCLE STARTS AT ___	OFF CYCLE LENGTH ___	REPEAT ___ TIMES

TABLE 20–5 Load shedding schedule.

Priority	Channel	kW draw	Max off minutes	Used for
1				
2				
3				
4				
5				
6				
7				
8				
9				
10				
11				
12				

Set point (0-999) ___   Deadband size (0-99) ___

TABLE 20–6 Reset boiler control schedule.

OUTPUT CHANNEL TO RESET HEATING (1-12) > ___
OFF AT (0-70) > ___
HEATING RATIO (4-9) > ___ TO 6
OUTSIDE AIR SENSOR (1-5) > ___

TABLE 20–7 Rest cooler control schedule.

OUTPUT CHANNEL TO RESET COOLING (1-12) > ___
OFF AT (0-70) > ___
COOLING RATIO (4-9) > ___ TO 6
OUTSIDE AIR SENSOR (1-5) > ___

■ JOB SHEET 20–3 ■

Name _____

Score _____ Date _____

PERFORMANCE OBJECTIVE
Run program and insert schedule data; troubleshoot and correct faults inserted by the instructor by isolating the various devices in the energy management system.

REFERENCE
Heating, Ventilating, and Air Conditioning Fundamentals (Chapter 20); TEC/CUBE program manual

MATERIAL
Computer, TEC/CUBE Energy Management Training Unit

JOB 20–3
Troubleshoot an energy management system

PROCEDURE
1. Observe the program flowchart prior to running the program.
2. Bring up the program on the computer.
3. Start with the flowchart on p. 306 and fill in the data.
4. Upon completion of the flowcharts, have instructor energize a fault control switch.
5. Isolate controls to find fault.
6. Call instructor to grade job sheet.

TEC/CUBE(R) TRAINING SYSTEM

PROGRAM FLOWCHART

These diagrams will show you where the program leads from each of the menus. The arrows point to the direction of the flow, and the boxes each represent a different screen.

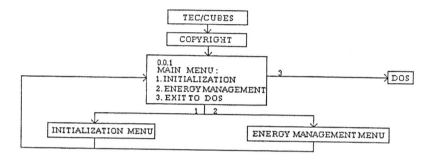

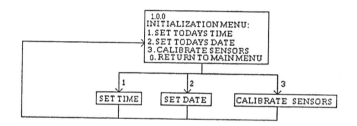

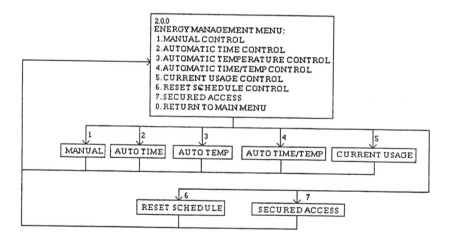

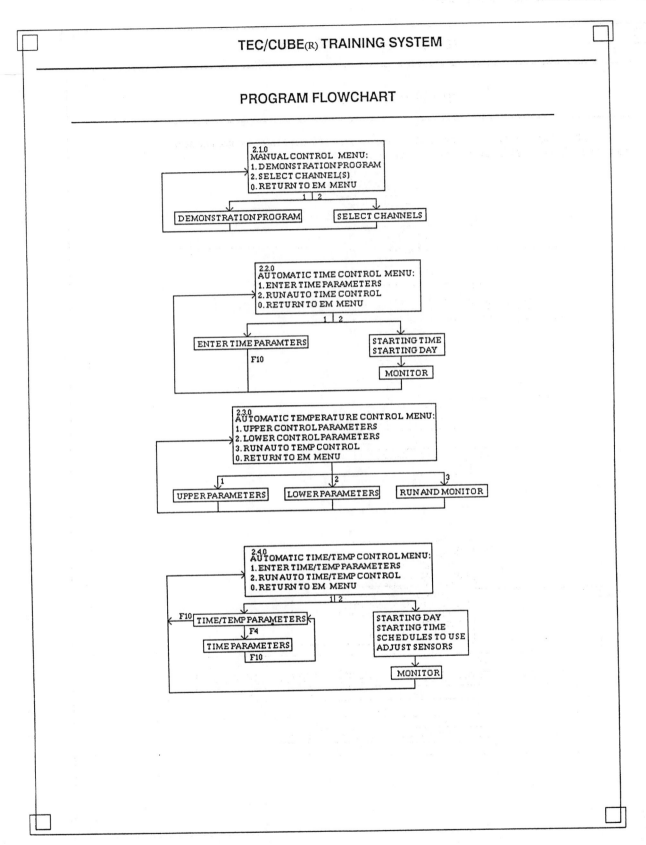

TEC/CUBE(R) TRAINING SYSTEM

PROGRAM FLOWCHART

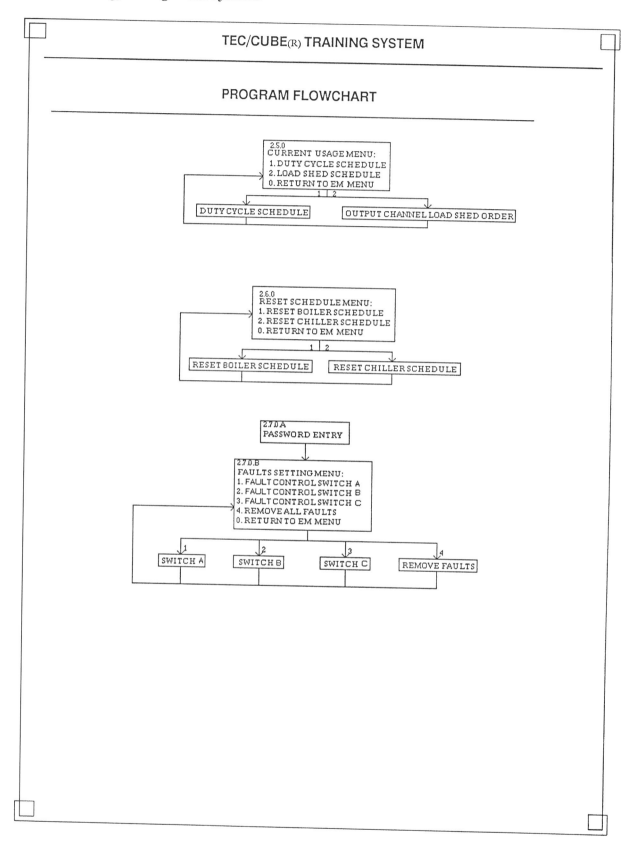

■ MULTIZONE INPUT/OUTPUT SUMMARY ■

Name _____

Score _____ Date _____

Input/Output Summary

System, Apparatus, or Area Point Description	Analog — Measured (Temperature, Pressure, RH, KW)	Analog — Calc. (KWH, Enthalpy, Run Time, Efficiency)	Binary (Status, Filter, Smoke, Freeze, Off-Slow-Fast, Hi-Lo)	Commandable — (OFF-ON, OFF-AUTO-ON)	Commandable — Pos. (Cntrl. Pt. Adj., Dmpr. Pos.)	Grad.	System Features — Alarms (Hi Analog, Low Analog, Hi Binary, Low Binary, Proof)	System Features — Programs (Time Scheduling, Demand Limiting, Duty Cycle, Start/Stop Opt., Enthalpy Opt., Reset, Event Program, DDC, Alarm Instruct, Maint. Work Order)	General (Intercom, Color Graphic)	Supplementary Notes
AHU #1										
SUPPLY FAN										
RETURN FAN										
RETURN AIR										
MIXED AIR										
DISCHARGE AIR										
FILTER										
ZONE PUBLIC										
ZONE - (7THUS)										
REHEAT										
RETURN AIR										
PREHEAT COIL										
COOLING COIL SUPPLY										
COOLING COIL RETURN										

Notes:

Page _____

Of _____

■ **FIGURE 20–3** Input/output summary. *(Control Systems International)*

■ WRITTEN EVALUATION ■

From the system apparatus or area point description column of the
above I/O chart, answer the following questions.

1. What points should be physically connected?
2. What points should be calculated through software?
3. What points should be displayed or printed?
4. What are the points consistent with system scope and return
   on investment?

APPENDIX

TEMP °F	113	141b	123	11	114	124	134a	12	500	22	502	AZ-50*	125	AZ-20**	13	23	503
−150.0								29.6	29.5	29.4	29.1	29.3	28.7	28.6	20.9	21.2	16.9
−140.0						29.7	29.6	29.4	29.2	29.1	28.5	28.9	28.1	27.9	16.8	17.1	11.1
−130.0						29.6	29.4	29.1	28.8	28.5	27.8	28.2	27.2	26.8	11.5	11.4	3.5
−120.0						29.5	29.1	28.6	28.3	27.7	26.7	27.3	25.9	25.3	4.5	3.9	3.1
−110.0					29.7	29.3	28.7	27.9	27.5	26.6	25.3	25.9	24.2	23.3	2.1	2.9	9.3
−100.0				29.7	29.5	29.0	28.0	27.0	26.9	25.1	23.3	23.9	21.8	20.5	7.6	9.0	16.9
−90.0				29.6	29.3	28.5	27.1	25.8	24.9	23.0	20.6	21.2	18.7	16.7	14.3	16.8	26.3
−80.0			29.7	29.5	29.0	27.8	25.7	24.1	22.9	20.2	17.2	17.6	14.7	11.9	22.5	26.3	37.7
−70.0	29.7	29.7	29.6	29.4	28.6	26.9	24.0	21.9	20.3	16.6	12.7	13.0	9.6	5.7	32.3	38.0	51.3
−60.0	29.7	29.5	29.4	29.1	28.0	25.7	21.6	19.0	17.0	11.9	7.2	7.0	3.1	1.1	43.9	52.0	67.3
−50.0	29.6	29.3	29.2	28.8	27.1	24.1	18.6	15.4	12.8	6.1	0.2	0.3	2.4	5.9	57.6	68.7	86.1
−40.0	29.4	29.0	28.8	28.3	26.1	22.0	14.7	11.0	7.6	0.6	4.1	4.8	7.3	11.8	73.3	88.4	107.8
−35.0	29.3	28.8	28.6	28.0	25.4	20.7	12.3	8.4	4.6	2.6	6.5	7.5	10.1	15.2	82.2	99.4	119.9
−30.0	29.2	28.6	28.3	27.7	24.7	19.3	9.7	5.5	1.2	4.9	9.2	10.4	13.2	18.9	91.6	111.3	132.8
−25.0	29.1	28.3	28.1	27.4	23.8	17.7	6.8	2.3	1.2	7.5	12.1	13.6	16.5	23.0	101.7	124.1	146.7
−20.0	29.0	28.1	27.7	26.9	22.9	15.9	3.6	0.6	3.2	10.2	15.3	17.0	20.2	27.5	112.5	137.8	161.4
−15.0	28.8	27.7	27.3	26.5	21.8	13.9	0.0	2.5	5.4	13.2	18.8	20.8	24.3	32.4	123.9	152.5	177.1
−10.0	28.6	27.3	26.9	25.9	20.6	11.6	2.0	4.5	7.8	16.5	22.6	25.0	28.6	37.8	136.1	168.2	193.9
−5.0	28.4	26.9	26.4	25.3	19.3	9.1	4.1	6.7	10.4	20.1	26.7	29.5	33.4	43.5	149.1	185.0	211.6
0.0	28.1	26.4	25.8	24.6	17.8	6.4	6.5	9.2	13.3	24.0	31.1	34.3	38.6	49.8	162.9	203.0	230.5
5.0	27.8	25.8	25.2	23.9	16.2	3.4	9.1	11.8	16.4	28.3	35.9	39.5	44.1	56.6	177.4	222.0	250.5
10.0	27.5	25.2	24.5	23.0	14.4	0.1	12.0	14.7	19.7	32.8	41.0	45.1	50.2	63.9	192.8	242.4	271.7
15.0	27.1	24.5	23.7	22.1	12.4	1.7	15.1	17.7	23.3	37.8	46.5	51.2	56.6	71.8	209.1	263.9	294.1
20.0	26.7	23.7	22.8	21.0	10.2	3.7	18.4	21.1	27.2	43.1	52.5	57.7	63.6	80.2	226.3	286.9	317.8
25.0	26.2	22.8	21.8	19.8	7.8	5.8	22.1	24.6	31.4	48.8	58.8	64.6	71.1	89.3	244.4	311.2	342.8
30.0	25.7	21.8	20.7	18.5	5.1	8.1	26.1	28.5	36.0	54.9	65.6	72.0	79.1	99.0	263.5	337.1	369.3
35.0	25.1	20.7	19.5	17.1	2.2	10.6	30.4	32.6	40.8	61.5	72.8	79.9	87.7	109.4	283.6	364.5	397.2
40.0	24.4	19.5	18.1	15.5	0.4	13.3	35.0	37.0	46.0	68.5	80.5	88.3	96.9	120.5	304.8	393.5	426.6
45.0	23.7	18.1	16.6	13.8	2.1	16.2	40.0	41.7	51.6	76.1	88.7	97.3	106.7	132.4	327.1	424.3	457.5
50.0	22.9	16.7	15.0	12.0	3.9	19.4	45.4	46.7	57.5	84.1	97.4	106.8	117.1	145.0	350.4	457.0	490.2
55.0	21.9	15.1	13.1	9.9	5.9	22.8	51.2	52.1	63.8	92.6	106.6	116.9	128.2	158.4	375.0	491.6	524.5
60.0	20.9	13.4	11.2	7.7	8.0	26.5	57.4	57.8	70.6	101.6	116.4	127.6	140.0	172.6	400.9	528.3	560.7
65.0	19.8	11.5	9.0	5.3	10.3	30.4	64.0	63.8	77.7	111.3	126.7	139.0	152.5	187.7	428.1	567.3	598.7
70.0	18.6	9.4	6.6	2.7	12.7	34.6	71.11	70.2	85.3	121.4	137.6	151.0	165.7	203.7	456.8	608.7	
75.0	17.2	7.2	4.1	0.1	15.3	39.1	78.6	77.0	93.4	132.2	149.1	163.7	179.7	220.6	487.2	652.7	
80.0	15.8	4.8	1.3	1.6	18.2	43.9	86.7	84.2	101.9	143.7	161.2	177.1	194.5	238.5	519.4		
85.0	14.2	2.3	0.9	3.2	21.2	49.0	95.2	91.7	110.9	155.7	174.0	191.3	210.2	257.4			
90.0	12.4	0.2	2.5	4.9	24.4	54.4	104.3	99.7	120.5	168.4	187.4	206.2	226.7	277.3			
95.0	10.5	1.7	4.2	6.8	27.8	60.2	113.9	108.2	130.5	181.8	201.4	222.0	244.1	298.4			
100.0	8.5	3.2	6.1	8.8	31.4	66.3	124.1	117.0	141.1	196.0	216.2	238.6	262.4	320.5			
105.0	6.2	4.8	8.1	10.9	35.3	72.8	134.9	126.4	152.2	210.8	231.7	256.1	281.6	343.8			
110.0	3.8	6.6	10.2	13.2	39.4	79.7	146.3	136.2	163.9	226.4	247.9	274.6	301.8	368.2			
115.0	1.2	8.4	12.6	15.7	43.8	87.0	158.4	146.5	176.3	242.8	264.9	294.0	323.1	393.9			
120.0	0.7	10.4	15.0	18.3	48.4	94.7	171.1	157.3	189.2	260.0	282.7	314.4	345.3	420.9			
125.0	2.2	12.4	17.7	21.1	53.3	102.8	184.5	168.6	202.7	278.1	301.3	335.9	368.7	449.2			
130.0	3.8	14.6	20.5	24.0	58.4	111.4	198.7	180.5	216.9	297.0	320.6	358.6	393.1	478.9			
135.0	5.5	16.9	23.5	27.1	63.9	120.4	213.5	192.9	231.8	316.7	341.2	382.4	418.6	510.0			
140.0	7.3	19.3	26.7	30.5	69.6	129.9	229.2	205.9	247.4	337.4	362.6	407.5	445.4	542.5			
145.0	9.2	21.8	30.2	34.0	75.6	139.9	245.6	219.5	263.7	359.1	384.9	433.9	473.3	576.5			
150.0	11.3	24.4	33.8	37.7	82.0	150.4	262.8	233.7	280.7	381.7	408.4	461.7	502.4	612.1			

TEMPERATURE: °F
PRESSURE: Psig
Vapor pressures are shown as psig.
Colored figures are shown as inches of mercury vacuum.

*AZ-50 is an azeotrope of 125/143a.
**AZ-20 is an azeotrope of 32/125.

A–1 Vapor pressures.

WeatherKing® Heating · Cooling RESIDENTIAL COOLING HEATING WORKSHEET

OWNER

NAME_____

STREET_____

CITY & STATE_____

LOCATION OF BUILDING

BUILDING NAME_____

STREET_____

CITY & STATE_____

BUILDING DATA REQUIRED

1. Building Dimensions, Length_____Ft. Width_____Ft. Exposed Wall Height_____Ft.

2. Area of Windows in each of four sides:
   Single Glass
   South_____Sq. Ft. West_____Sq. Ft. North_____Sq. Ft. East_____Sq. Ft.

   Double Glass
   South_____Sq. Ft. West_____Sq. Ft. North_____Sq. Ft. East_____Sq. Ft.

3. Area of Doors_____Sq. Ft. with Storm Door,_____Sq. Ft. without Storm Door

4. Sun Shading:
   Trees, Buildings, Awnings, Patio, Etc_____Blinds_____Shades_____Drapes

5. Construction--Exposed Walls _____

   Exposed Ceilings_____

   Unconditioned Partitions _____

6. Type of Roof & Ceiling _____Attic Space_____Flat-No Attic_____Cathedral

7. Insulation Thickness--Exposed Walls _____Inches, Exposed Ceiling _____Inches,
   Floor_____Inches.

8. No. of Bedrooms_____

9. Basement Wall--Above Grade_____ Below Grade _____

EQUIPMENT SELECTION SUMMARY

1. WeatherKing Furnace _____BTU Output_____
2. WeatherKing Package Air Conditioner_____BTU Output_____
3. WeatherKing Condensing Unit_____BTU Output_____
4. WeatherKing Evaporator_____
5. WeatherKing Coil Cabinet_____
6. WeatherKing Refrigerant Lines_____
7. WeatherKing Thermostat_____

Design Temp Difference Summer_____ Winter_____

REMARKS
_____
_____
_____
_____
_____
_____
_____
_____
_____
_____
_____

A–2 WeatherKing residential cooling/heating worksheets.
(Addison Products Company)

EXISTING INSTALLATION DATA

1. Heating System Type _____

2. Make of Furnace _____

3. BTUH Output _____

4. Existing Ductwork and System Operation Comments: _____

_____
_____
_____
_____
_____

NEW INSTALLATION DATA

1. Equipment Location:
   A. Unit: Basement _____ Utility Room _____ Attic _____

               Crawl Space _____ Roof _____ Outdoors _____

               Location of Condensing Unit _____

2. Controls:
   A. Thermostat: Heating Only _____ Heating-Cooling _____

                      Cooling Only _____ Other _____

   B. Humidistat: Duct Type _____ Wall Type _____

3. Utilities:
   A. Electrical Service: Capacity _____ Volts _____

                             Adequate _____

   B. Condensate Disposal: _____

   C. Gas Service: Type: _____ Meter Location _____

   D. Oil Service: Type: _____ Tank Location _____

   E. Chimney: Type: _____ Flue Size _____

                   Adequate _____

CONTRACTOR: _____

ADDRESS: _____

SURVEY BY: _____

ESTIMATE BY: _____

DATE: _____

ADDISON PRODUCTS COMPANY

ADDISON PRODUCTS COMPANY
Addison, Michigan
(517) 547-6131

DEARBORN STOVE COMPANY DIV.
Dallas, Texas
(214) 278-6161

WEATHERKING, INC.
Orlando, Florida
(305) 894-2891

Litho U. S. A.   SA-75-33

A–2 *(Continued)*

OUTDOOR DESIGN CONDITIONS—SUMMER AND WINTER

STATE AND CITY	NORMAL DESIGN COND.—SUMMER July at 3:00 PM			AVG. DAILY RANGE	MAXIMUM DESIGN COND.—SUMMER July at 3:00 PM			NORMAL DESIGN COND. WINTER		WIND DATA Avg. Velocity and Prevailing Direction		Elevation Above Sea Level (ft)	Latitude (deg)
	Dry-Bulb (F)	Wet-Bulb (F)	Moisture Content* (gr/lb of dry air)	Dry-Bulb (F)	Dry-Bulb (F)	Wet-Bulb (F)	Moisture Content† (gr/lb of dry air)	Dry-Bulb (F)	Annual Degree Days	Summer	Winter		
ALABAMA													
Anniston	95	78	117.5	19				5	2806			733	34
Birmingham	95	78	117.5	19	99			10	2611	5.0 S	8.0 N	694	34
Mobile	95	80	131	12	95	82	155.6	15	1566	9.0 SW	9.9 N	10	31
Montgomery	95	78	117.5	15				10	2071		7.5 NW	293	32
ARIZONA													
Flagstaff	90	65	81	26	90			−10	7242		7.7 SW	6,894	35
Phoenix	105	76	94	30	113	78	126.9	25	1441	5.0 W	5.4 E	1,108	33
Tucson	105	72	77	30				25		5.0 W	5.2 NW	2,376	32
Winslow	100	70	85					−10				4,853	35
Yuma	110	78	93	30				30	1036		6.7 N	146	33
ARKANSAS													
Fort Smith	95	76	104.5	16	103			10	3226	7.0 E	8.3 E	448	35
Little Rock	95	78	117.5	16	103	83	145.5	5	3009	6.0 NW	8.3 NW	324	35
CALIFORNIA													
Bakersfield	105	70	54	25				25				499	35
El Centro	110	78	94									43	33
Eureka	90	65	52		110	75	95.9	30	4758	7.0 N	7.3	132	41
Fresno	105	74	76	35				25	2403	8.0 NW	5.4 NW	287	37
Laguna Beach				9	82	70	103.0					10	34
Long Beach	90	70	78	14								47	34
Los Angeles	90	70	78	14	94			35	1391	6.0 SW	6.4 NE	261	34
Oakland	85	65	60	17	94	68	99.3	30				17	38
Montague								0				2,635	42
Pasadena	95	70	70										34
Red Bluff	100	70	62									305	40
Sacramento	100	72	73	18				30	2680		7.2 SE	116	39
San Bernadino	105	72	65					35	1596	7.0 W	6.3 NW	26	34
San Diego	85	68	75	10	88	74	78.4	35					33
San Francisco	85	65	60	17				35	3137	12.0 W	7.5 N	17	38
San Jose	91	70	76.5					25	2823			100	37
Williams				40	110	80	74.4					86	39
COLORADO													
Denver	95	64	60	25	99	68	89.4	−10	5839	7.0 S	7.5 S	5,221	40
Durango	95	65	70									6,558	37
Fort Collins								−30					41
Grand Junction	95	65	62	24	102	68	86.2	−15	5613	6.0 SE	4.4 NW	4,587	39
Pueblo	95	65	63	25				−20	5558		7.9 NW	4,770	38
CONNECTICUT													
Bridgeport	95	75	99	14				0				9	41
Hartford	93	75	102	16	94	82		0	6113	7.0 S	8.7 NW	58	42
New Haven	95	75	99	14	95			0	5880	7.0 S	9.4 N	23	41
Waterbury								−15					42
DELAWARE													
Wilmington	95	78	117.5	15				0		10.0 SW	NW	134	40
DIST. OF COLUMBIA													
Washington	95	78	117.5	18	99	84	155.6	0	4561	5.0 S	7.8 NW	72	39
FLORIDA													
Apalachicola	95	80	131					25	1252	5.0 SW	8.4	23	30
Jacksonville	95	78	117.5	17	99	82	150.5	25	1185	8.0 SW	9.0 NE	18	30
Key West	98	78	112.5					45	59	9.0 SE	10.6 NE	23	25
Miami	91	79	131	12	92	81	150.5	35	185	7.0 SE	10.1 E	11	26
Pensacola	95	78	117.5	12				20	1281		10.9 N	408	31
Tampa	95	78	117.5	14	95			30	571	6.0 NE	8.6 NE	25	28
Tallahassee								25	1463		N	68	30

*Corresponds to dry-bulb and wet-bulb temperatures listed, and is corrected for altitude of city.

†Corresponds to peak dewpoint temperature, corrected for altitude.

A–3 Outdoor design conditions—summer and winter. (*From Handbook of Air Conditioning System Design, Carrier Corp. Courtesy McGraw-Hill Book Company.*)

OUTDOOR DESIGN CONDITIONS—SUMMER AND WINTER (CONT.)

STATE AND CITY	NORMAL DESIGN COND.—SUMMER July at 3:00 PM			AVG. DAILY RANGE	MAXIMUM DESIGN COND.—SUMMER July at 3:00 PM				NORMAL DESIGN COND. WINTER		WIND DATA Avg. Velocity and Prevailing Direction		Eleva-tion Above Sea Level (ft)	Lati-tude (deg)
	Dry-Bulb (F)	Wet-Bulb (F)	Moisture Content* (gr/lb of dry air)	Dry-Bulb (F)	Dry-Bulb (F)	Wet-Bulb (F)	Moisture Content† (gr/lb of dry air)		Dry-Bulb (F)	Annual Degree Days	Summer	Winter		
GEORGIA														
Atlanta	95	76	109.5	18	101	82	150.5		10	2985	7.0 NW	11.7 NW	975	34
Augusta	98	76	100	18					10	2306		6.5 NW	195	34
Brunswick	95	78	117.5											31
Columbus	98	76	100											33
Macon	95	78	117.5	18					15	2338	5.0 S	6.7 NW	408	33
Savannah	95	78	117.5	17	99				20	1635	8.0 SW	9.5 NW	42	32
IDAHO														
Boise	95	65	54.5	31	109	71	92.6		—10	5678	5.0 NW	9.1 SE	2,705	44
Lewiston	95	65	44	28					5	5109		4.1 E	763	46
Pocatello	95	65	61	28	100				—5	6741		8.9 SE	4,468	43
Twin Falls									—10			W		42
ILLINOIS														
Cairo	98	78	112.5						0	3957		9.8	319	37
Chicago	95	75	99	19	104	80	140.6		—10	6282	10.0 NE	12.0 SW	594	42
Danville									—5			NW		40
Moline	96	76	103	22	103	83	155.6		—10				594	41
Peoria	96	76	103	20	100				—10	6004	8.0 S	8.3 S	602	41
Springfield	98	77	106	20					—10	5446		11.9 NW	603	40
INDIANA														
Evansville	95	78	117.5	19	102	82	150.5		0	4410	7.0 SW	9.7 S	388	38
Fort Wayne	95	75	99	20	100				—10	6232	8.0 SW	10.4 SW	777	41
Indianapolis	95	76	104.5	18	99				—10	5458	9.0 SW	11.3 S	715	40
South Bend									—5			SW	773	42
Terre Haute	95	78	124										1,146	40
IOWA														
Cedar Rapids									—5					42
Davenport	95	78	117.5	18					—15	6252		10.5 NW	648	42
Des Moines	95	78	123	18	102				—15	6375	6.0 SW	10.1 NW	800	42
Dubuque	95	78	117.5						—20	6820		7.1	740	43
Fort Dodge									—20					42
Keokuk	95	78	117.5						—10	5663		8.2 SW	637	41
Sioux City	95	78	124	19	102				—20	6905	10.0 S	11.5 NW	1,111	43
Waterloo									—15					43
KANSAS														
Concordia	95	78	125	20					—10	5425		7.7 S	1,425	39
Dodge City	95	78	132	21	106				—10	5069		10.6	2,522	38
Salina					111				—15			NW	1,226	39
Topeka	100	78	109.5	19					—10	5075	10.0 S	9.2 S	991	39
Wichita	100	75	98	21	110	79	126.9		—10	4644	11.0 S	12.4 S	1,300	38
KENTUCKY														
Lexington									0	4792		13.3 SW	989	38
Louisville	95	78	117.5	22	99				0	4417	7.0 SW	9.8 SW	459	38
LOUISIANA														
Alexandria									20			N	89	32
New Orleans	95	80	131	13	95	83	161.2		20	1203	6.0 SW	8.6 N	9	30
Shreveport	100	78	109.5	15	102	83	150.5		20	2132	5.0 S	8.8 SE	197	33
MAINE														
Augusta	90	73	95	13										45
Bangor	90	73	95	13									362	45
Bar Harbor									—15			NW		44
Belfast									—5					44
Eastport	90	70	78	13					—10	8445	7.0 S	12.6 W	100	45
Millinocket									—15					46
Presque Isle										9644		NW		47
Portland	90	73	95	13	93				—5	7377	7.0 S	10.4 NW	47	44
Rumford									—20					44

*Corresponds to dry-bulb and wet-bulb temperatures listed, and is corrected for altitude of city.
†Corresponds to peak dewpoint temperature, corrected for altitude.

A–3   *(Continued)*

OUTDOOR DESIGN CONDITIONS—SUMMER AND WINTER (CONT.)

STATE AND CITY	NORMAL DESIGN COND.—SUMMER July at 3:00 PM			AVG. DAILY RANGE	MAXIMUM DESIGN COND.—SUMMER July at 3:00 PM			NORMAL DESIGN COND. WINTER		WIND DATA Avg. Velocity and Prevailing Direction		Elevation Above Sea Level (ft)	Latitude (deg)
	Dry-Bulb (F)	Wet-Bulb (F)	Moisture Content* (gr/lb of dry air)	Dry-Bulb (F)	Dry-Bulb (F)	Wet-Bulb (F)	Moisture Content† (gr/lb of dry air)	Dry-Bulb (F)	Annual Degree Days	Summer	Winter		
MARYLAND													
Baltimore	95	78	117.5	18	99			0	4487	6.0 SW	8.2 NW	14	39
Cambridge								5			NW		39
Cumberland	95	75	99	18									39
Frederick								−5			NW		40
Frostburg								−5			W		40
Salisbury								10			NW		40
MASSACHUSETTS													
Amherst								−10			NW		42
Boston	92	75	104	13	96	78	135.9	0	5936	9.0 SW	12.4 W	14	42
Fall River								−10					42
Fitchburg	93	75	102	17				−10	6743	W	NW	402	43
Lowell								−15					43
Nantucket	95	75	99					0			14.8	45	41
New Bedford								0					42
Plymouth								−5			W		42
Springfield	93	75	102	17				−10		9.0 SW		199	42
Worcester	93	75	102	17				0				625	42
MICHIGAN													
Alpena	95	75	99					−10	8278		11.0 SW	615	45
Big Rapids								−15			NW		43
Detroit	95	75	99	19	101	79	135.9	−10	6560	10.0 SW	12.0 SW	619	42
Escanaba								−15	8777		9.5 NW		46
Flint	95	75	99	20				−10		W	W	766	43
Grand Rapids	95	75	99	20	98			−10	6702	8.0 W	12.1 NW	638	43
Kalamazoo								−5			W		42
Lansing	95	75	104	20				−10	7149		9.8 SW	861	43
Ludington								−10	7458		11.9 W	652	44
Marquette	93	73	90	20	96			−10	8745		10.6 NW		47
Saginaw	95	75	99									601	43
Sault Ste Marie								−20	9307		8.9 SE	724	47
MINNESOTA													
Alexandria								−25			NW		
Duluth	93	73	96	19				−25	9723	13.4 SW	13.4 SW	1,128	47
Minneapolis	95	75	103	17	102			−20	7966	10.0 S	11.3 NW	839	45
St. Cloud								−25					46
St. Paul	95	75	99	17	103	79	131.1	−20	7975	8.0 SE	9.5 NW	719	45
MISSISSIPPI													
Jackson				21	103	83	155.6	15		5.0 SW	7.7 SE	316	32
Meridian	95	79	124	21				10	2330	4.0 SW	6.3 N	410	32
Vicksburg	95	78	117.5	21	96			10	2069	6.0 SW	8.3	226	32
MISSOURI													
Columbia	100	78	109.5	19				−10	5070		8.9 SW	739	39
Kansas City	100	76	106.5	19	109	79	135.9	−10	4962	9.0 S	10.3 NW	741	39
Kirksville				19	108	82	150.5				SW	969	40
St. Louis	95	78	117.5	20	108	81	135.9	0	4596	9.0 S	11.8 S	465	39
St. Joseph								−10	5596		9.3 NW	817	40
Springfield				18	98	79	135.9	−10	4569	8.0 S	10.9 SE	1,301	37
MONTANA													
Billings	90	66	70	20	104			−25	7213		12.4 W	3,119	46
Butte								−20			NW	5,538	46
Great Falls								−20			SW	3,687	48
Havre	95	70	82	20				−30	8416	7.0 E	9.4 SW	2,498	49
Helena	95	67	71	20	97	70	77.4	−20	7930	7.0 SW	7.4 SW	4,090	47
Kalispell	95	65	56					−20	8032		5.2	3,004	48
Miles City								−35	7591		5.6 S	2,629	47
Missoula	95	66	49	20				−20	7604		E	3,205	47

*Corresponds to dry-bulb and wet-bulb temperatures listed, and is corrected for altitude of city.

†Corresponds to peak dewpoint temperature, corrected for altitude.

A–3 *(Continued)*

OUTDOOR DESIGN CONDITIONS—SUMMER AND WINTER (CONT.)

STATE AND CITY	NORMAL DESIGN COND.—SUMMER July at 3:00 PM			AVG. DAILY RANGE	MAXIMUM DESIGN COND.—SUMMER July at 3:00 PM			NORMAL DESIGN COND. WINTER		WIND DATA Avg. Velocity and Prevailing Direction		Elevation Above Sea Level (ft)	Latitud (deg)
	Dry-Bulb (F)	Wet-Bulb (F)	Moisture Content* (gr/lb of dry air)	Dry-Bulb (F)	Dry-Bulb (F)	Wet-Bulb (F)	Moisture Content† (gr/lb of dry air)	Dry-Bulb (F)	Annual Degree Days	Summer	Winter		
NEBRASKA													
Grand Island								−20				1,856	41
Lincoln	95	78	124	20	106			−10	5980	9.0 S	10.6 S	1,180	41
Norfolk								−15		NW	NW		42
North Platte	95	78	135	26	104	76	74.4	−20	6384	6.0 S	7.9 W	2,805	41
Omaha	95	78	123	20	108	80	131.1	−10	6095	8.0 S	9.7 NW	978	41
Valentine	95	78	135	20				−25	7197		9.2	2,627	43
York								−15			NW		
NEVADA													
Las Vegas	115	75	76	40				20			S	1,882	36
Reno	95	65	62	41	102	66	66.9	−5	5621	7.0 SW	6.0 W	4,493	40
Tonopah								5	5812		9.9 SE	5,421	38
Winnemucca	95	65	62	40				−15	6357	7.0 SW	8.1 NE	4,293	42
NEW HAMPSHIRE													
Berlin								−25					45
Concord	90	73	95	14				−15	7400	5.0 NW	6.2 NW	289	43
Keene								−20			NW		43
Manchester	90	73	95	14	92							171	43
Portsmouth	90	73	95	14									43
NEW JERSEY													
Atlantic City	95	78	117.5	14				5	5015	13.0 SW	15.8 NW	8	39
Bloomfield	95	75	99	14								125	41
Camden				14	102	82	145.5	0		10.0 SW		30	40
East Orange	95	75	99	14								173	41
Jersey City	95	75	99					0			NW		41
Newark	95	75	99	14	99	81	140.6	0	5500	13.0 SW	17.1 NW	10	41
Paterson	95	75	99	14	95					13.0 SW		10	41
Sandy Hook								0	5369		16.1		41
Trenton	95	78	117.5	14	96			0	5256	9.0 SW	10.9 NW	56	40
NEW MEXICO													
Albuquerque	95	70	94.5	26	98	68	95.9	0	4517	8.0 SW	7.3 N	5,101	35
Roswell	95	70	87	25				−10	3578	6.0 S	7.1 S	3,643	32
Santa Fe	90	65	80	30	90			0	6123	6.0 SE	7.1 NE	7,000	36
NEW YORK													
Albany	93	75	102	18	97	78	131.1	−10	6648	7.0 S	10.5 S	19	43
Binghamton	95	75	103.5					−10	6818		6.8 NW	915	42
Buffalo	93	73	90	18	93	77	126.9	−5	6925	12.0 SW	17.1 W	604	43
Canton	90	73	95					−25	8305	8.0	10.5	458	43
Cortland								−10			NW		43
Glens Falls								−15			W		43
Ithaca								−15	6914		11.3 NW		42
Jamestown								−10			SW		42
Lake Placid								−20			W		44
New York City	95	75	99	14	100	81	145.5	0	5280	13.0 S	16.8 NW	10	41
Ogdensburg								−20			SW		45
Oneonta								−15			SW		43
Oswego	93	73	90					−10	7186		12.1 S	363	43
Rochester	95	75	102	18	95			−5	6772	8.0 SW	9.6 W	543	43
Schenectady	93	75	102	18								235	43
Syracuse	93	75	102	18	96			−10	6899	9.0 S	11.2 S	400	43
Watertown								−15			SW		44
NORTH CAROLINA													
Asheville	93	75	114.5	19	93			0	4236	6.0 NW	9.5 NW	2,192	36
Charlotte	95	78	117.5	16				10	3224	5.0 SW	7.3 SW	809	35
Greensboro	95	78	123.5	15				10	3849		7.9 SW	896	37
Raleigh	95	78	117.5	15	98	82	155.6	10	3275	6.0 SW	7.9 SW	345	36
Wilmington	95	78	117.5	15	95	81	150.5	15	2420	7.0 SW	9.4 SW	6	34

*Corresponds to dry-bulb and wet-bulb temperatures listed, and is corrected for altitude of city.

†Corresponds to peak dewpoint temperature, corrected for altitude.

A–3 *(Continued)*

OUTDOOR DESIGN CONDITIONS—SUMMER AND WINTER (CONT.)

STATE AND CITY	NORMAL DESIGN COND.—SUMMER July at 3:00 PM			AVG. DAILY RANGE	MAXIMUM DESIGN COND.—SUMMER July at 3:00 PM			NORMAL DESIGN COND. WINTER		WIND DATA Avg. Velocity and Prevailing Direction		Elevation Above Sea Level (ft)	Latitude (deg)
	Dry-Bulb (F)	Wet-Bulb (F)	Moisture Content* (gr/lb of dry air)	Dry-Bulb (F)	Dry-Bulb (F)	Wet-Bulb (F)	Moisture Content† (gr/lb of dry air)	Dry-Bulb (F)	Annual Degree Days	Summer	Winter		
NORTH DAKOTA													
Bismarck	95	73	95.5	19	103			−30	8937	9.0 NW	9.1 NW	1,670	47
Devils Lake	95	70	77					−30	10104		10.1 W	1,481	48
Fargo	95	75	104.5	19				−25			10.9 NW	900	47
Grand Forks								−25	9871		NW	832	48
Williston	95	73	96.5					−35	9301	8.0 SE	8.6 W	1,919	48
OHIO													
Akron	95	75	99	19				−5		7.0 SW	8.5 SW	104	41
Cincinnati	95	78	117.5	22	106	81	145.5	0	4990	11.0 S	14.7 SW	553	39
Cleveland	95	75	99	19	101	79	135.9	0	6144			651	42
Columbus	95	76	104.5	23	95			−10	5506	9.0 SW	11.6 SW	724	40
Dayton	95	78	123	23	99			0	5412	8.0 SW	11.1 SW	900	40
Lima								−5					41
Sandusky	95	75	99					0	6095		11.0	608	42
Toledo	95	75	99	19	99			−10	6269	10.0 SW	12.1 SW	589	42
Youngstown	95	75	99	19								1,186	41
OKLAHOMA													
Ardmore								10			N	762	34
Bartlesville								−10			N		37
Oklahoma City	101	77	108	21	104			0	3670	10.0 S	11.5 S	1,254	35
Tulsa	101	77	101.5		106	79	140.6	0		10.0 S	N	804	36
OREGON													
Baker	90	66	71	19				−5	7197		5.6 SE	3,501	44
Eugene	90	68	67	19				−15				366	44
Medford	95	70	76	19								1,428	42
Pendleton								−15			W	1,494	46
Portland	90	68	67	19	99	70	103.0	10	4353	6.0 NW	7.3 S	30	46
Roseburg	90	66	57	19						4.0 N		523	42
Wamic								0			W		45
PENNSYLVANIA													
Altoona	95	75	99	14				−5				1,469	40
Bethlehem								−5					41
Erie	93	75	102	18				−5	6363	9.0 S	13.6 SW	670	42
Harrisburg	95	75	99	14				0	5412		7.6 NW	339	40
New Castle								0			NW		41
Oil City	95	75	99	18									42
Philadelphia	95	78	117.5	14	97			0	4739	10.0 SW	11.0 NW	26	40
Pittsburgh	95	75	105	14	98	79	126.9	0	5430	9.0 SW	11.6 W	1,248	40
Reading	95	75	99					0	5232		9.0	311	40
Scranton	95	75	99	14	95			−5	6218	6.0 SW	7.6 SW	746	41
Warren								−15			NW		41
Williamsport								−5			NW	525	42
RHODE ISLAND													
Block Island	95	75	99						5897		20.6 NW	46	41
Pawtucket	93	75	102	14									41
Providence	93	75	102	14				0	5984	10.0 NW	12.1 NW	8	42
SOUTH CAROLINA													
Charleston	95	78	117.5	17	98	82	155.6	15	1866	10.0 SW	10.5 SW	9	33
Columbia	95	75	99	17				10	2488		8.0 SW	401	34
Greenville	95	76	104.5	17				10	3059	7.0 NE	8.4	982	35
SOUTH DAKOTA													
Huron	95	75	106	19	106	76	126.9	−20	7940	10.0 SE	10.7 NW	1,282	44
Rapid City	95	70	85	22	103	71	95.9	−20	7197	7.0 W	8.0 W	3,231	44
Sioux Falls	95	75	99	20				−20			NW	1,427	43

*Corresponds to dry-bulb and wet-bulb temperatures listed, and is corrected for altitude of city.
†Corresponds to peak dewpoint temperature, corrected for altitude.

A–3 (Continued)

OUTDOOR DESIGN CONDITIONS—SUMMER AND WINTER (CONT.)

STATE AND CITY	NORMAL DESIGN COND.—SUMMER July at 3:00 PM			AVG. DAILY RANGE	MAXIMUM DESIGN COND.—SUMMER July at 3:00 PM			NORMAL DESIGN COND. WINTER		WIND DATA Avg. Velocity and Prevailing Direction		Elevation Above Sea Level (ft)	Latitude (deg)
	Dry-Bulb (F)	Wet-Bulb (F)	Moisture Content* (gr/lb of dry air)	Dry-Bulb (F)	Dry-Bulb (F)	Wet-Bulb (F)	Moisture Content† (gr/lb of dry air)	Dry-Bulb (F)	Annual Degree Days	Summer	Winter		
TENNESSEE													
Chattanooga	95	76	104.5	18	98			10	3238	6.0 SW	7.7 NW	689	35
Johnson City								0			W		36
Knoxville	95	75	103.5	17	100	79	135.9	0	3658	6.0 SW	7.2 SW	921	36
Memphis	95	78	117.5	18	103	83	155.6	0	3090	7.0 SW	9.3 W	271	35
Nashville	95	78	117.5	17	98			0	3613	8.0 W	9.8 NW	485	36
TEXAS													
Abilene	100	74	93					15	2573	9.0 S	10.1 S	1,748	32
Amarillo	100	72	91.6	22	101	75	110.4	−10	4196	11.0 S	12.1 SW	3,657	35
Austin	100	78	109.5	19				20	1679		8.3 N	625	31
Brownsville	95	80	131	20	96	80	150.5	30	628	9.0 SE	10.4 SE	35	26
Corpus Christi	95	80	131					20	965	13.0 SE	11.0 SE	21	28
Dallas	100	78	109.5	21	105	80	135.9	0	2367	8.0 S	10.6 NW	460	33
Del Rio	100	78	115					15	1501	10.0 SE	8.0 SE	1,020	29
El Paso	100	69	73	23	101	72	106.6	10	2532	9.0 E	9.0 NW	3,720	32
Fort Worth	100	78	109.5	21				10	2355	10.0	10.5 NW	708	33
Galveston	95	80	131	14				20	1174	9.0 S	11.2 SE	6	29
Houston	95	80	131	14	100	81	150.5	20	1315	8.0 S	10.5 SE	52	30
Palestine	100	78	109.5					15	2068		8.0	555	32
Port Arthur	95	79	124					20	1532		10.7	64	30
San Antonio	100	78	109.5	19	102	83	166.4	20	1435	7.0 SE	8.3 NE	646	29
UTAH													
Modena	95	65	66	25	97	66	80.3	−15	6598	11.0 SW	9.0	5,479	38
Logan								−15					42
Ogden								−10			S	4,446	41
Salt Lake City	95	65	61	25	102	68	89.4	−10	5650	7.0 S	7.8 SE	4,222	41
VERMONT													
Bennington								−10					43
Burlington	90	73	95	17	91			−10	8051	8.0 S	11.6 S	308	44
Rutland	90	73	95	17				−20					43
VIRGINIA													
Cape Henry	95	78	117.5					10	3538		14.0	24	37
Lynchburg	95	75	99	16	99			5	4068		8.1	386	37
Norfolk	95	78	117.5	16	95			15	3364	11.0 S	12.1 N	11	37
Richmond	95	78	117.5	16	98			15	3922	6.0 SW	8.1 SW	162	38
Roanoke	95	76	111.5	16				0	4075		8.2 W	1,194	38
WASHINGTON													
North Head	85	65	60					20	5367		16.1	199	
Seattle	85	65	60	17	86	70	99.3	15	4815	7.0 N	9.8 SE	14	48
Spokane	93	65	54.5	28	106	68	71.9	−15	6318	7.0 SW	6.2 SW	1,879	48
Tacoma	85	64	55.5	17				15	5039		8.0	279	47
Tatoosh Island								15	5857		18.9	110	48
Walla Walla	95	65	47.5	28	105			−10	4910		5.4 S	952	46
Wenatchee	90	65	52	20									48
Yakima	95	65	48	20				5	5585		4.1	1,160	47
WEST VIRGINIA													
Bluefield	95	75	99	16									37
Charleston	95	75	99	16	102			0		4.0 SW	W	603	38
Elkins								−10	5800		6.2 W	2,006	39
Huntington	95	76	104.5	16				−5			W		38
Martinsburg								−5				540	39
Parkersburg	95	75	99	16	98			−10	4928	4.0 SE	7.2 SW	615	39
Wheeling	95	75	99	14				−5					40

*Corresponds to dry-bulb and wet-bulb temperatures listed, and is corrected for altitude of city.
†Corresponds to peak dewpoint temperature, corrected for altitude.

A–3 (Continued)

OUTDOOR DESIGN CONDITIONS—SUMMER AND WINTER (CONT.)

STATE AND CITY	NORMAL DESIGN COND.—SUMMER July at 3:00 PM			AVG. DAILY RANGE	MAXIMUM DESIGN COND.—SUMMER July at 3:00 PM			NORMAL DESIGN COND. WINTER		WIND DATA Avg. Velocity and Prevailing Direction		Elevation Above Sea Level (ft)	Latitude (deg)
	Dry-Bulb (F)	Wet-Bulb (F)	Moisture Content* (gr/lb of dry air)	Dry-Bulb (F)	Dry-Bulb (F)	Wet-Bulb (F)	Moisture Content† (gr/lb of dry air)	Dry-Bulb (F)	Annual Degree Days	Summer	Winter		
WISCONSIN													
Ashland								−20			SW	885	42
Eau Claire								−20			NW		45
Green Bay	95	75	99	14	99	79	131.1	−20	7931	8.0 S	10.5 SW	589	45
La Crosse	95	75	99	17	100	83	161.2	−25	7421	6.0 S	9.3 S	673	44
Madison	95	75	103.5	18	96			−15	7405	8.0 SW	10.1 NW	938	43
Milwaukee	95	75	99	14	99			−15	7079	9.0 SW	12.1 W	619	43
WYOMING													
Casper								−20			SW	5,321	43
Cheyenne	95	65	68.5	28				−15	7536	9.0 S	13.3 NW	6,139	42
Lander	95	65	66	28				−18	8243	5.0 SW	3.9	5,448	44
Sheridan					102			−30	7239	5.0 NW	4.9 NW	3,773	45
CANADA													
PROVINCE AND CITY													
ALBERTA													
Calgary	90	66	71					−29	9520	9.7	10.1	3,540	51
Edmonton	90	68	77					−33	10320	8.9	7.6	2,219	54
Grand Prairie								−39			7.9	2,190	55
Lethbridge								−32	8650		15.0	3,018	50
McMurray								−42				1,216	57
Medicine Hat	90	65						−35	8650	9.1	9.0	2,365	50
BRITISH COLUMBIA													
Estevan Point								17			9.9	20	49
Fort Nelson								−38			3.7	1,230	59
Penticton								−6				1,121	50
Prince George								−32	9500		7.2	2,218	54
Prince Rupert								8	6910		8.0	170	54
Vancouver	80	67	78					11	5230		7.7	22	49
Victoria								15	5410		12.3	228	48
MANITOBA													
Brandon								−32	10930			1,200	50
Churchill								−42	16810		14.7	115	59
The Pas								−39			6.4	894	54
Winnipeg	90	71	83.5					−29	10630	11.5	12.0	786	50
NEW BRUNSWICK													
Campbellton								−11				42	48
Fredericton	90	75	107					−6	8830		9.2	164	46
Moncton								−8	8700		14.9	248	46
Saint John								−3	8380	7.9	13.8	119	45
NEWFOUNDLAND													
Corner Brook								−1	9210			40	49
Gander								−3	9440		17.2	482	49
Goose Bay								−26	12140		10.3	144	53
Saint Johns								1	8780		19.3	463	48
NORTHWEST TERRITORIES													
Aklavik								−46	17870			30	68
Fort Norman								−42	16020			300	65
Frobisher								−47				68	
Resolute								−42			9.2	56	
Yellowknife								−47				682	62

*Corresponds to dry-bulb and wet-bulb temperatures listed, and is corrected for altitude of city.
†Corresponds to peak dewpoint temperature, corrected for altitude.

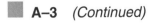 **A–3**   *(Continued)*

OUTDOOR DESIGN CONDITIONS—SUMMER AND WINTER (CONT.)

CANADA PROVINCE AND CITY	NORMAL DESIGN COND.—SUMMER July at 3:00 PM			AVG. DAILY RANGE	MAXIMUM DESIGN COND.—SUMMER July at 3:00 PM			NORMAL DESIGN COND. WINTER		WIND DATA Avg. Velocity and Prevailing Direction		Elevation Above Sea Level	Latitude
	Dry-Bulb (F)	Wet-Bulb (F)	Moisture Content* (gr/lb of dry air)	Dry-Bulb (F)	Dry-Bulb (F)	Wet-Bulb (F)	Moisture Content† (gr/lb of dry air)	Dry-Bulb (F)	Annual Degree Days	Summer	Winter	(ft)	(deg)
NOVA SCOTIA													
Halifax	90	75	107					4	7570	6.6	9.6	83	45
Sydney								1	8220	9.9	13.1	197	46
Yarmouth								7	7520		13.5	136	44
ONTARIO													
Fort William								−24	10350	8.4	9.6	644	48
Hamilton								0	6890			303	43
Kapuskasing								−30	11790		10.0	752	49
Kingston								−11	7810		10.8	340	44
Kitchener								−3	7380			1,100	43
London								−1			11.9	912	43
North Bay								−20		9.6	11.3	1,210	46
Ottawa	90	75	107					−15	8830	8.9	11.1	339	45
Peterborough								−11				648	44
Souix Lookout								−33			8.5	1,227	50
Sudbury								−17				837	47
Timmins								−26				1,100	48
Toronto	93	75	102					0	7020	8.1	14.1	379	43
Windsor								3			12.3	637	42
Sault Ste. Marie	93	75	102									635	47
PRINCE EDWARD ISLAND													
Charlottetown								−3	8380	8.7	11.3	74	46
QUEBEC													
Arvida								−19	10440		8.2	375	
Knob Lake								−40				1,605	55
Mont Joli								−11			13.3	150	48
Montreal	90	75	107					−9	8130	9.9	12.3	187	46
Port Harrison								−39			13.4	66	58
Quebec City	90	75	107					−12	9070	9.0	12.4	296	47
Seven Islands								−20				190	50
Sherbrooke								−12	8610		8.2	620	45
Three Rivers								−13				50	46
SASKATCHEWAN													
Prince Albert								−41	11430		4.9	1,414	53
Regina	90	71	92.5					−34	10770	12.4	12.1	1,884	50
Saskatoon	90	70	81					−37	10960	10.7	9.7	1,645	52
Swift Current								−33	9660		14.6	2,677	50
YUKON TERRITORY													
Dawson								−56	15040			1,062	64
Whitehorse								−43			8.7	2,289	61

*Corresponds to dry-bulb and wet-bulb temperatures listed, and is corrected for altitude of city.
†Corresponds to peak dewpoint temperature, corrected for altitude.

A–3 (Continued)

COOLING ESTIMATE WORKSHEET

AREA SQ. FT. (X) BTU MULT. FACTOR = BTU/HR.

	AREA SQ. FT.	BTU MULT. FACTORS INSULATION						BTU/HR
		0"	1"	2"	3"	4"	6"	
NET EXPOSED WALLS & PARTITIONS								
8" MASONRY BLOCK PLASTERED	_____	10.6						_____
8" MASONRY BLOCK FURRED & PLASTERED	_____	7.5	3	2.5	2			_____
8" SOLID CONCRETE	_____	16.2	5.3					_____
FRAME - BRICK VEN. STUCCO, WOOD OR ALUMINUM SIDING	_____	9.7	5	4.5	3.5			_____
CEILINGS - UNDER								
VENTED ATTIC OR UNCONDITIONED SPACE	_____	11.6	5	3.5	3	2.3	2	_____
ROOF CEILING COMBINATION, NO ATTIC	_____	15.6		4.8		3.5	2.6	_____
FLOORS - OVER								
UNCONDITIONED ROOM	_____	4	3	2.2	1.6	1.3	1	_____
VENTED CRAWL SPACE OR GARAGE	_____	5.5	3.6	2.4	1.8	1.4	1	_____
CLOSED CRAWL SPACE, BASEMENT OR CONCRETE SLAB ON GROUND	_____	0	0	0	0	0	0	_____

	AREA SQ. FT.	UNSHADED		DRAPES OR BLINDS		BTU/HR
GLASS AREA - INCLUDING GLASS DOORS		SINGLE	STORM	SINGLE	STORM	
NORTH	_____	46	33	33	26	_____
EAST & WEST	_____	117	97	78	65	_____
SOUTH	_____	65	52	46	34	_____

	AREA SQ. FT.	SINGLE	STORM	BTU/HR
ALL DIRECTIONS IF SHADED BY PATIO, AWNING OR OTHER OUTSIDE SHADE	_____	40	26	_____

	AREA SQ. FT.		BTU/HR
DOORS - OTHER THAN GLASS	_____	X 17.5	_____
OCCUPANCY LOAD - NO. OF BEDROOMS	_____	X 920	_____
APPLIANCE LOAD - AVERAGE HOME	_____		1600

OUTSIDE DESIGN TEMP. FACTOR

85° = .50	90° = .75	95° = 1.0
100° = 1.25	105° = 1.50	119° = 1.75

DUCTWORK FACTOR
DUCT LOCATED INSIDE CONDITIONED SPACE   1.0

DUCT LOCATED OUTSIDE CONDITIONED SPACE
WITH 1" BLANKET INSULATION   1.15
With 2" BLANKET INSULATION   1.1

TOTAL BTU/HR. HEAT GAIN FOR
20° DESIGN TEMP. DIFFERENCE = _____
DUCT FACTOR   X   ABOVE = _____

DESIGN TEMP. FACTOR
X ABOVE IS TOTAL BTU/HR   = _____

NOTES
1. LATENT LOAD INCLUDED IN ABOVE
2. VENTILATION INCLUDED IN ABOVE
3. NET WALL EQUAL GROSS WALL LESS WINDOWS
    AND DOORS.

A–4 Cooling estimate worksheet.

HEATING ESTIMATE WORKSHEET

AREA SQ. FT. (X) BTU MULT. FACTOR = BTU/HR.

	AREA SQ. FT.	BTU MULT. FACTORS INSULATION						BTU/HR
NET EXPOSED WALLS & PARTITIONS		0"	1"	2"	3"	4"	6"	
8" MASONRY BLOCK PLASTERED	_____	48						_____
8" MASONRY BLOCK FURRED & PLASTERED	_____	30	13					_____
8" SOLID CONCRETE	_____	67						_____
FRAME - BRICK VEN. STUCCO, WOOD OR ALUMINUM SIDING	_____	28	11	9	7			_____
WALLS BELOW GRADE	_____	6						_____
CEILINGS - UNDER								
VENTED ATTIC OR UNCONDITIONED SPACE	_____	60	14	12	8	7	5	_____
ROOF CEILING COMBINATION, NO ATTIC	_____	31	11.5	9	7	6.2	5	_____
FLOORS - OVER								
OVER UNCONDITIONED ROOM	_____	14		7	5.5	4.6	3.4	_____
OVER VENTED CRAWL SPACE OR GARAGE	_____	28		9	7	5.6	4	_____
BASEMENT FLOOR	_____	3						_____
CONCRETE SLAB FLOOR ON GROUND LIN. FT.	_____	75	58	50				_____
OVER HEATED SPACE	_____	0	0	0	0	0	0	_____
GLASS AREA		SINGLE			STORM			
DOUBLE HUNG, SLIDING OR CASEMENT	_____	150			90			_____
FIXED OR PICTURE	_____	140			85			_____
JALOUISE	_____	750			220			_____
SLIDING GLASS DOOR	_____	250			200			_____
DOORS								
NO WEATHER STRIPPING OR STORM	_____			450				_____
WITH WEATHER STRIPPING OR STORM	_____			240				_____
WITH WEATHER STRIPPING AND STORM	_____			130				_____
VENTILATION								
NO. OF BEDROOMS	_____		X		220			_____

OUTSIDE DESIGN TEMP. DIFFERENCE FACTOR

50° = .50	60° = .60	70° = .70
80° = .80	90° = .90	100° = 1.00

DUCTWORK FACTOR

DUCTWORK LOCATED INSIDE CONDITIONED SPACE 1.0
DUCTWORK LOCATED OUTSIDE CONDITIONED SPACE
WITH 1" DUCT WRAP   1.15
WITH 2" DUCT WRAP   1.1

TOTAL BTU/HR HEAT LOSS FOR 100 DESIGN TEMP. DIFF.   _____

DESIGN TEMP. FACTOR _____ X ABOVE   _____

DUCT FACTOR _____ X ABOVE EQUALS
TOTAL BTU/HR LOSS   _____

A–5 Heating estimate worksheet.